Praise for *Brains*

"The '*Brains*' in Dale Purves' book are both the focus of his research and the intellectual giants with whom he mixed in his remarkable career. At one level *Brains* is a charming autobiography; at another a vivid personal account of 50 years in the evolution of neuroscience. But it also lays down a challenge to the mainstream view that simple, sequential analysis of nerve cells and their responses can explain the apparently impossible task of seeing the world as it really is on the basis of the infinitely ambiguous retinal image. *Brains* is a delight—for its insights into both the scientists and the science of the brain."

—**Colin Blakemore**, Universities of Oxford and Warwick

"Dale Purves has been a leading figure in brain science for 40 years: He has made numerous discoveries, founded departments, and written major textbooks. In *Brains*, he tells the story of his scientific journey and intertwines it in an elegant and accessible way with two others: the uneven progress of the field over the past half century and the new view of brain function to which the field's shortcomings have led him. Some neuroscientists are likely to disagree with Purves' heretical theories, but none can afford to ignore them."

—**Joshua R. Sanes**, Harvard University

"*Brains* is much more than a book about brains. It is a journey that takes the reader through the modern history of neurobiology, a personal account that illuminates both what we know about brains and the mysteries that remain in understanding how brains work."

—**Terrence J. Sejnowski**, Howard Hughes Medical Institute, Salk Institute, and University of California at San Diego

"Dale Purves had the good fortune to be present at the birth of the new discipline of neuroscience, which condensed in the 1960s and 1970s from elements of physiology, anatomy, and neurology. The first half of the book tracks his odyssey through two of the linchpin departments of the new discipline, those at Harvard Medical School and University College London, to introduce the basic principles of neuroscience for nonspecialists. The book's second half encapsulates Purves' major contribution to the field of neuroscience, and particularly to visual neuroscience. The experimental tools developed during the past half century provide us with an excellent picture of the activity of individual neurons. At the same time, these tools seduce neuroscientists into a reductionist approach that focuses on the microscopic details of visual analysis. Perhaps owing to his training in philosophy, Purves recognized the evolutionary importance of the way that we see: The brain does not report an objective reality but instead provides its best guess at that reality on the basis of the fragmentary information at hand. The brain can be misled by unusual and contrived visual inputs—the basis of visual 'illusions'—but this scarcely represents a defect: The brain interprets patterns of light, darkness, and color on the basis of the most plausible natural stimuli, and it remains for researchers to learn how this interpretation comes about. *Brains* offers a guide to thinking like a neuroscientist."

—**A. James Hudspeth**, Rockefeller University

"This is a lucid, easy-to-read summary that is fascinating reading for anyone interested in what we know and do not know about how brains work. Purves brings together a unique expertise and priceless personal observations about several subfields of brain research and the scientists who have shaped our present understanding of it over the past eventful fifty years."

—**Pasko Rakic**, Yale University School of Medicine

"Dale Purves' *Brains* is my favorite sort of reading—an engaging and intelligent scientific autobiography full of vivid personal and historical accounts; the story not only of a life but of an intellectual pursuit. Purves has a unique voice, lively, outspoken, and very human—and his love of science comes through on every page."

—**Oliver Sacks**

"*Brains* is an engaging tour of human neuroscience from one of its most distinguished and opinionated practitioners. Dale Purves is a lively and informative guide to the field, having been at the scene of some of its great discoveries and having made many important discoveries himself."

—**Steven Pinker**, Harvard University, author of *The Stuff of Thought*

"A rare account of both the modern history of key discoveries in brain research by someone who was there and responsible for many of them and also a heartfelt account of the joy of it all. Dale Purves has given us an inside view of a life in science and explains with clarity what it all means."

—**Michael S. Gazzaniga**, University of California, Santa Barbara, author of *Human: The Science Behind What Makes Us Unique*

"*Brains* is a delightful book that weaves together Dale Purves' personal neuroscience history with the history and current status of the field. I enjoyed it start to finish."

—**Joseph LeDoux**, New York University, author of *The Emotional Brain* and *Synaptic Self*

"This book is many things. It's the memoir of an immensely likeable human (who I only previously knew as a distant giant in my field). It's people with strong personalities that give lie to the notion that science is an affectless process. But most of all, it is a clear, accessible, affectionate biography of neuroscience. This is a terrific book."

—**Robert M. Sapolsky**, Stanford University, author of *Why Zebras Don't Get Ulcers*

"Both highly entertaining and educational. A masterpiece."

—**Bert Sakmann**, Max Planck Institute for Medical Research, winner of the Nobel Prize in Physiology or Medicine

Brains

Brains

How They Seem to Work

Dale Purves

Vice President, Publisher: Tim Moore
Associate Publisher and Director of Marketing: Amy Neidlinger
Acquisitions Editor: Kirk Jensen
Editorial Assistant: Pamela Boland
Operations Manager: Gina Kanouse
Senior Marketing Manager: Julie Phifer
Publicity Manager: Laura Czaja
Assistant Marketing Manager: Megan Colvin
Cover Designer: Sandra Schroeder
Managing Editor: Kristy Hart
Project Editor: Betsy Harris
Copy Editor: Krista Hansing Editorial Services, Inc.
Proofreader: Kathy Ruiz
Senior Indexer: Cheryl Lenser
Compositor: Jake McFarland
Manufacturing Buyer: Dan Uhrig

FT Press offers excellent discounts on this book when ordered in quantity for bulk purchases
or special sales. For more information, please contact U.S. Corporate and Government Sales,
1-800-382-3419, corpsales@pearsontechgroup.com. For sales outside the U.S., please contact
International Sales at international@pearson.com.

ISBN-10: 0-13-705509-9
ISBN-13: 978-0-13-705509-8

Pearson Education LTD.
Pearson Education Australia PTY, Limited
Pearson Education Singapore, Pte. Ltd.
Pearson Education North Asia, Ltd.
Pearson Education Canada, Ltd.
Pearson Educación de Mexico, S.A. de C.V.
Pearson Education—Japan
Pearson Education Malaysia, Pte. Ltd.

Library of Congress Cataloging-in-Publication Data
Purves, Dale.
Brains: how they seem to work/Dale Purves.
    p. cm.
Includes index.
ISBN 978-0-13-705509-8 (hardback: alk. paper)   1. Brain—Popular works.
2. Neurobiology—Popular works.   I. Title.
QP376.P868 2010
612.8'2—dc22
                                        2009048550

For Shannon, who has always
deserved a serious explanation.

Contents

Preface xi

Chapter 1   Neuroscience circa 1960 1

Chapter 2   Neurobiology at Harvard 17

Chapter 3   Biophysics at University College 37

Chapter 4   Nerve cells versus brain systems 51

Chapter 5   Neural development 69

Chapter 6   Exploring brain systems 87

Chapter 7   The visual system: Hubel and
Wiesel redux 105

Chapter 8   Visual perception 123

Chapter 9   Perceiving color 143

Chapter 10   The organization of perceptual qualities 161

Chapter 11   Perceiving geometry 179

Chapter 12   Perceiving motion 201

Chapter 13   How brains seem to work 219

Suggested reading 235

Glossary 241

Illustration credits 275

Acknowledgments 281

About the author 283

Index 285

Preface

This book is about the ongoing effort to understand how brains work. Given the way events determine what any scientist does and thinks, an account of this sort must inevitably be personal (and, to a greater or lesser degree, biased). What follows is a narrative about the ideas that have seemed to me especially pertinent to this hard problem over the last 50 years. And although this book is about brains as such, it is also about individuals who, from my perspective, have significantly influenced how neuroscientists think about brains. The ambiguity of the title is intentional.

The idiosyncrasies of my own trajectory notwithstanding, the story reflects what I take to be the experience of many neuroscientists in my generation. Thomas Kuhn, the philosopher of science, famously distinguished the pursuit of what he called "normal science" from the more substantial course corrections that occur periodically. In normal science, Kuhn argued, scientists proceed by filling in details within a broadly agreed-upon scheme about how some aspect of nature works. At some point, however, the scheme begins to show flaws. When the flaws can no longer be patched over, the interested parties begin to consider other ways of looking at the problem. This seems to me an apt description of what has been happening in brain science over the last couple decades; in Kuhn's terms, this might be thought of as a period of grappling with an incipient paradigm shift. Whether this turns out to be so is for future historians of science to decide, but there is not much doubt that those of us interested in the brain and how it works have been struggling with the conventional wisdom of the mid- to late twentieth century. We are looking hard for a better conception of what brains are trying to do and how they do it.

I was lucky enough to have arrived as a student at Harvard Medical School in 1960, when the first department of neurobiology in the United States was beginning to take shape. Although I had no way of knowing then, this contingent of neuroscientists, their mentors, the colleagues they interacted with, and their intellectual progeny provided much of the driving force for the rapid advance of neuroscience over this period and for many of the key ideas about the brain that are now being questioned.

My interactions with these people as a neophyte physician convinced me that trying to understand what makes us tick by studying the nervous system was a better intellectual fit than pursuing clinical medicine. Like every other neuroscientist of my era, I set out learning the established facts in neuroscience, getting to know the major figures in the field, and eventually extending an understanding of the nervous system in modest ways within the accepted framework. Of course, all this is essential to getting a job, winning financial support, publishing papers, and attaining some standing in the community. But as time went by, the ideas and theories I was taught about how brains work began to seem less coherent, leading me and others to begin exploring alternatives.

Although I have written the book for a general audience, it is nonetheless a serious treatment of a complex subject, and getting the gist of it entails some work. The justification for making the effort is that what neuroscientists eventually conclude about how brains work will determine how we humans understand ourselves. The questions being asked—and the answers that are gradually emerging—should be of interest to anyone inclined to think about our place in the natural order of things.

Dale Purves
Durham, NC
January 2010

1

Neuroscience circa 1960

My story about the effort to make some sense of the human brain begins in 1960, not long after I arrived in Boston to start my first year at Harvard Medical School. Within a few months, I began learning the foundations of brain science (as it was then understood) from a remarkable group of individuals who had themselves only recently arrived at Harvard and were mostly not much older than me.

The senior member of the contingent was Stephen Kuffler, then in his early 50s and already a central figure in twentieth-century neuroscience. Otto Krayer, the head of the Department of Pharmacology at the medical school, had recruited Kuffler to Harvard from Johns Hopkins only a year earlier. Kuffler's mandate was to form a new group in Pharmacology by hiring faculty whose interests spanned the physiology, anatomy, and biochemistry of the nervous system. Until then, Harvard had been teaching neural function as part of physiology, brain structure as a component of traditional anatomy, and brain chemistry as aspects of pharmacology and biochemistry.

Kuffler had presciently promoted to faculty status two postdoctoral fellows who had been working with him at Hopkins: David Hubel and Torsten Wiesel, then 34 and 36, respectively. He'd also hired David Potter and Ed Furshpan, two even younger neuroscientists who had recently finished fellowships in Bernard Katz's lab at University College London. The last of his initial recruits was Ed Kravitz, who, at 31, had just received his Ph.D. in biochemistry from the University of Michigan. This group (Figure 1.1) made up the Department of Neurobiology in 1966, which soon became a standard departmental category in U.S. medical schools as the field burgeoned both intellectually and as a magnet for research funds. In the

Figure 1.1 The faculty Steve Kuffler recruited when he came to Harvard in 1959. Clockwise from the upper left: Ed Furshpan, Steve Kuffler, David Hubel, Torsten Wiesel, Ed Kravitz, and David Potter. This picture was taken in 1966, about the time the pharmacology group became a department in its own right under Kuffler's leadership. (Courtesy of Jack McMahan)

neuroscience course medical students took during my first year, Furshpan and Potter taught us how nerve cell signaling worked, Kravitz taught us neurochemistry, and Hubel and Wiesel taught us about the organization of the brain (or, at least, the visual part of it, which was their bailiwick). Kuffler gave a pro forma lecture or two, but this sort of presentation was not his strong suit, and he had the good sense and self-confidence to let these excellent teachers carry the load.

For me, and for most of my fellow first-year medical students, this instruction was being written on a blank slate. I had graduated the previous June from Yale as a premed student majoring in philosophy, and my background in hard science was minimal. Then, as now, premeds were required to take courses in only general chemistry, organic chemistry, biology, and physics. The premed course in biology at Yale that I took in 1957 was antediluvian, consisting of a first semester of botany in which we pondered the differences between palmate and pinnate leaves, and a more useful but nonetheless mundane second semester on animal physiology. We learned little about modern genetics, although James Watson and Francis Crick had discovered the structure of DNA several years earlier and the revolution in molecular biology was underway. John Trinkaus, the young Yale embryologist who taught us, ensured his popularity with the all-male class by well practiced off-color jokes that would today be grounds for dismissal.

Since the age of 14 or 15, I was determined to be a doctor. I decided in college that psychiatry was a specialty that would combine medicine with my interest in philosophy of the mind (the idea that the nuts and bolts of the brain biology might be involved in all this did not loom large in my thinking). In my senior year at Yale, I had been one of a dozen members of the Scholars of the House program that permitted us to forego formal course requirements and spend our time writing a full-blown thesis on a subject of our choosing (or the equivalent—two members of the group were aspiring novelists, and one was a poet). I am somewhat embarrassed to say that my thesis was on Freud as an existentialist. Although I enjoyed the perks of the program, the main lesson I learned from writing this philosophical treatise was that thinking about mental functions without the tools needed to rise above the level of speculation was frustrating and likely to be a waste of time. Therefore, in early spring 1961, I was especially attuned to what I might get out of our first-year medical school course on the nervous system. I assumed it would be the beginning of a new effort to learn about the brain in a more serious way than I had managed as an undergraduate, and so it was.

Even the least interested among us paid close attention to the distillation by Kuffler's young faculty of the best thinking about the nervous system that had emerged during the preceding few decades. The major topics they covered were the cellular structure of the nervous system, the electrochemical mechanisms nerve cells use to convey information over long distances, the means by which nerve cells communicate this information to other neurons at synapses, the biochemistry of the neurotransmitters underlying this communication, and, finally, the overall organization of the brain and what little was known about its functional properties. A truism often heard in science education is that much of what one learns will change radically in the near future. In fact, the fundamentals of neuroscience that we were taught in spring 1961 would, with a few important updates, serve reasonably well today.

The part of the course that was easiest to absorb concerned the cellular structure of the brain and the rest of the nervous system. A long-standing debate in the late nineteenth century focused on whether the cells that comprise the nervous system were separate entities or formed a syncytium in which, unlike the cells in other organs, protoplasmic strands directly connected these elements to form a continuous network. In light of the apparently special operation of the brain, the idea of a protoplasmic network was more sensible then than it seems now. The microscopes of that era were not good enough to resolve this issue by direct observation, and the discreteness of neurons (*neurons* and *nerve cells* are synonyms) was not definitively established until the advent of electron microscopy in neuroscience in the early 1950s.

The warring parties in this debate were Spanish neuroanatomist Santiago Ramon y Cajál, who favored the ultimately correct idea that individual cells signaled to one another by special means at synapses, and the equally accomplished Italian physician and scientist Camillo Golgi, who argued that a network made more sense. Ironically, Cajál won the day by using a staining technique that Golgi had invented, showing that the neurons absorbed the stain as individual elements (Figure 1.2A). Their joint contributions to understanding neuronal structure were enormous, and they shared the Nobel Prize for Physiology or Medicine in 1906. This work led to an increasingly deep understanding of the diversity and detail of nerve cell structure established by the legion of neuroanatomists that followed.

In addition to the individuality of nerve cells, a second key feature of neuronal anatomy is structural polarization (Figure 1.2B). Neurons generally have a single process extending from the cell body, called the *axon*, that conveys information to other nerve cells (or to non-neural target cells such as muscle fibers and gland cells), and a second set of more complex branches called *dendrites*, which receive information from the axonal endings of other nerve cells. Together with newly acquired electron microscopical evidence about neuronal structure (Figure 1.2C), this comprised a fundamental part of what we learned in 1961.

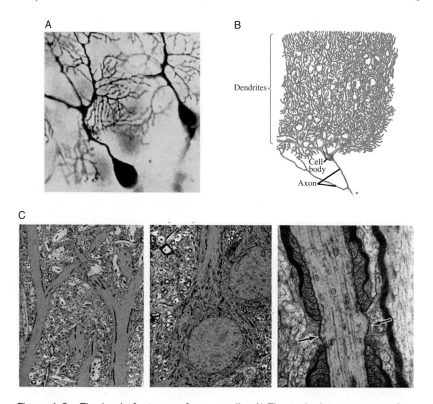

Figure 1.2   The basic features of nerve cells. A) The typical appearance of a nerve cell revealed by Golgi's silver stain—the method Cajál used to demonstrate neuronal individuality. B) Diagrammatic representation of the same class of neurons as in (A), showing the relationship of the cell body, dendrites, and the axon (the asterisk indicates that the axon travels much farther than shown). C) Electron micrographs from the work of Sanford Palay, another teacher who had been recruited to the Department of Anatomy at Harvard Medical School in 1961. The micrographs show the same elements as (B), but at the much higher magnification possible with this method. The panel on the left shows dendrites (purple), the central panel shows two-cell bodies and their nuclei, and the right panel shows part of an axon (blue). (*The Fine Structure of the Nervous System: Neurons and Their Supporting Cells, 3e,* by Alan Peters, Sanford L. Palay, Henry Webster; © Oxford University Press, 1991. Reprinted by permission of Oxford University Press, Inc.)

The second body of information we learned was how nerve cells transmit electrical signals and communicate with one another. For much of the first half of the twentieth century, neuroscientists sought to understand how neurons convey a signal over axonal processes that, in humans, can extend up to a meter, and how the information carried by the axon is passed on to the nerve cells (or other cell types) it contacts. By 1960, both of these processes were pretty well understood. Potter and Furshpan, Kuffler's young recruits fresh from their postdoctoral training in London, shared the duty of teaching us how axons conduct an electrical signal—the *action potential*, as it had long been called (Figure 1.3). Although efforts to understand the action potential can be traced back to Luigi Galvani and Alessandro Volta's studies of animal electricity in the late eighteenth century, British physiologists Alan Hodgkin and Andrew Huxley had only recently provided a definitive understanding in work that they had published in 1952. Galvani discovered that an electrical charge applied to nerves caused muscles to twitch, although he misunderstood the underlying events; it was Volta who showed that electricity produced by the battery he invented triggered this effect. By the mid-nineteenth century, German physician and physiologist Emil du Bois-Reymond and others had shown that an electrical disturbance was progressively conducted along the length of axons; by the beginning to the twentieth century, it was apparent that this process depended on the sodium and potassium ions in the fluid that normally bathes the inside and outside of neurons and all other cells. Despite these advances, the way action potentials are normally generated remained unclear; by the 1930s, it was obvious that the most salient problem in neuroscience at the time was to understand the action potential mechanism and how this electrical signal is conducted along axons. After all, this signaling process is the basis of all brain functions.

Hodgkin had begun working on the action potential problem in 1937 as a fellow at Rockefeller University and the Marine Biological Laboratory in Woods Hole on Cape Cod. Back at Cambridge in 1938, he and Huxley, who was one of his students, began to collaborate. Their joint effort was interrupted by the war, but by the late 1940s, Hodgkin and Huxley had shown on the basis of a beautiful set of observations using the giant axon of the squid that the mechanism of the action potential was a voltage-dependent opening of ion channels, or "pores," in the nerve cell membrane that enabled sodium ions to rush into an axon, causing the spike in the membrane voltage illustrated in Figure 1.3 (the technical advantages of this very large axon explains why they used squid, as well as the reason for working at Woods Hole and, subsequently, at the Plymouth Marine Laboratory in the United Kingdom). They found that the voltage reduction across the nerve cell membrane that caused the inrush of sodium underlying the spike was the advancing action potential itself. By successively depolarizing

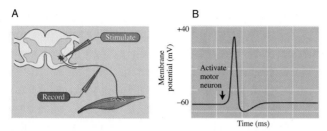

Figure 1.3   The electrical signal (the action potential or spike) that is conducted along nerve cell axons, transferring information from one place to
another in the nervous system. A) Diagram of a cross-section of the spinal
cord showing the stimulation of a spinal motor neuron (red) whose axon
extends to a muscle. The electrical disturbance conducted along the axon is
recorded with another electrode. B) The brief disturbance elicited in this way
can be monitored on an oscilloscope, a sensitive voltmeter that can measure
changes across a nerve cell membrane over time. The action potential lasts
only a millisecond or two at the point of recording, traveling down the axon to
the muscle at a rate of about 50 meters per second. If, as in this case, the
action potential is being recorded with an electrode placed inside the axon, it
is apparent that the phenomenon entails a brief change in the membrane
potential that first depolarizes the axon (the upward part of the trace), and
then quickly repolarizes the nerve cell to its resting level of membrane voltage
(the downward part of the trace that returns to the baseline). (After Purves,
Augustine, et al., 2008)

each little segment of an axon, the electrical disturbance (the spike) is
conducted progressively from one end of the axon to the other. A good
way to visualize this process is how the burning point travels along the
length a fuse, as seen in old western movies. The analogy is that the
heat at the burning point ignites the powder in the next segment of the
fuse, enabling conduction along its length. Hodgkin and Huxley's discovery of the action potential mechanism was quickly recognized as a
major advance; ten years after the work had been published, the Nobel
committee awarded them the 1963 prize in Physiology or Medicine.

This much deeper understanding of the mechanism of the action
potential in the late 1940s and early 1950s made an equally important
question more pressing: When the signal reaches the end of an axon,
how is the information then conveyed to the target cell or cells? For
example, how does the information carried by the axon of the motor
neuron in Figure 1.3A cause the muscle fibers it contacts to contract?
Teaching us the answer to this question about neural signaling also
fell to Potter and Furshpan, and their zeal was missionary. Although
the class included about 110 of us, they called each of us by name

within a week. They had just finished fellowships working together in Katz's lab, and it was Katz who supplied the answer to this question about neural signaling. Trained in medicine in Leipzig in the early 1930s, Katz, who was Jewish, had emigrated to England in 1935. Archibald Hill, then head of the Department of Biophysics at University College London and a preeminent figure in the field of energy and metabolism, particularly as these issues pertained to muscles, took Katz under his wing. Katz had worked on related problems as a medical student; the two had corresponded, so it was natural for Hill to take him on as a graduate student, even though the political importance of his sponsorship left Katz deeply indebted (a bronze bust of "A.V.," as Hill was known to his faculty, had a prominent place in Katz's office in the department he eventually inherited from Hill).

His work on muscle energetics with Hill led Katz to a growing interest in the signaling between nerve and muscle. When he finished his doctoral work in 1938, Katz went to Australia as a Carnegie Fellow to work more directly on this aspect of signaling with John Eccles, the leading neurophysiologist then working on synaptic transmission. While in Australia, Katz became a British citizen, served in the Royal Air Force as a radar officer, married, and eventually returned to University College London as Hill's assistant director in 1946. Back in England, he briefly collaborated with Hodgkin and Huxley on understanding the action potential and coauthored a paper with them that reported one of the major steps in this work in 1948. Katz had joined the effort at the end of Hodgkin and Huxley's remarkable collaboration, and had the foresight to see that he would be better off working on a related but different problem: how action potentials convey their effects to target cells by means of synapses. As fellows in Katz's lab, Furshpan and Potter had been involved in one aspect of this work just before Kuffler recruited them to Harvard in 1959.

The concept of synaptic transmission had emerged from Cajál's demonstration that nerve cells are individual entities related by contiguity instead of continuity. British physiologist Charles Sherrington coined the term *synapse* in 1897 (the Greek word *synapsis* means "to clasp"), which was soon adopted, although not until the 1950s were synapses seen directly by means of electron microscopy. The information that Potter and Furshpan conveyed to us in their lectures was essentially what Katz had discovered during the first decade or so of his work on the nerve-muscle junction. The steps in this process are illustrated in Figure 1.4.

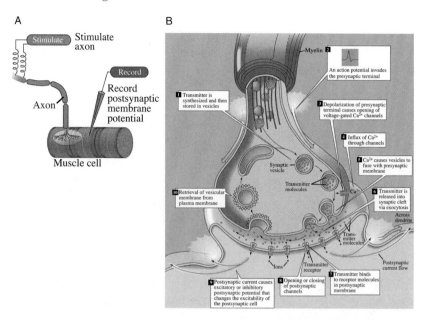

Figure 1.4 Synapses and synaptic transmission. A) The neuromuscular junction, the prototypical synapse studied by Katz and his collaborators as a means of unraveling the basic mechanisms underlying the process of chemical synaptic transmission. B) Diagram of a generic chemical synapse, illustrating the steps in the release of a synaptic transmitter agent from synaptic vesicles, which is stimulated by the arrival of an action potential at the axon terminal. The binding of the transmitter molecules to receptors embedded in the membrane of the postsynaptic cell enables the signal to carry forward by its effect on the membrane of the target cell, either exciting it to generate another action potential or inhibiting it from doing so. (After Purves, Augustine, et al., 2008)

Katz's work on synapses had effectively begun during his fellowship with Eccles in Sydney in the late 1930s. Eccles, an Australian, had trained at Oxford as a Rhodes Scholar under Sherrington, and had earned his doctoral degree there in 1929. When Katz arrived in Australia in 1938, Eccles had been working on synaptic transmission, having been interested in this problem since his time with Sherrington. Another recent arrival in Sydney was Kuffler, who was at that point just another bright young Jewish emigré. He had come in 1937 after completing medical training in Hungary, and had been working in the pathology department of the Medical School in Sydney. Kuffler met Eccles by chance on the tennis court. During their conversations, he expressed openness to doing something more interesting than his work in pathology, and Eccles eventually invited him to join his lab as a fellow. Katz and Kuffler quickly became friends and, with Eccles' blessing, began working on how action potentials activated muscle fibers (Figure 1.5).

A    B

Figure 1.5   A) John Eccles (center) with his two protégés, Kuffler (on the left) and Katz, in Sydney in 1941. B) The same trio at a meeting at Oxford in 1972 (Eccles is on the right). (Courtesy of Marion Hunt)

Eccles had long thought that synaptic transmission depended on the direct passage of electrical current from the axonal endings to target cells, although circumstantial evidence already suggested that axon endings released a chemical transmitter agent. This indirect evidence for chemical transmission came largely from the physiological and biochemical studies of John Langley at Cambridge and his student Henry Dale working on the peripheral autonomic nervous system early in the twentieth century, and from Otto Loewi working on the neural control of heart muscle in the 1920s. Despite Eccles's contrary view, Katz and Kuffler provided strong evidence for the chemical nature of neuromuscular transmission during their collaboration in Sydney. In later years, neither Katz nor Kuffler expressed much affection for Eccles, who could be overbearing in the prosecution of his ideas.

When Katz returned to University College London in 1946 as Hill's assistant director, he began a series of studies showing that the arrival of the action at axon terminals causes the release of transmitter molecules stored in small membrane-bound units called *synaptic vesicles* (Figure 1.4B). In the presence of sufficient calcium ions, the vesicles fuse with the terminal membrane and release neurotransmitter molecules into the narrow synaptic cleft. Katz and his collaborators eventually showed that calcium ions enter the axon ending through another type of membrane channel specific to axon terminals that the depolarization of nerve terminals opens when the action potential arrives. The transmitter molecules then diffuse across the synaptic cleft and bind to receptor proteins in the membrane of the target cell, an interaction that either opens or closes ion channels associated with the receptors. The net effect of various ions moving through these transmitter-activated channels is to help trigger an action potential in the target cell (synaptic excitation) or to prevent an action potential from occurring (synaptic inhibition).

Katz had chosen to work on the neuromuscular junction instead of synapses in the central nervous system (the choice Eccles had made) because of their simplicity and accessibility. Although synaptic transmission in muscle causes the muscle fibers to contract (in ways that Huxley was then beginning to work out in the 1950s in a lab not far from Katz's at University College), no one really doubted that the same mechanisms operated at synapses in the brain and the rest of the nervous system to pass on information—an idea that has been amply confirmed. Although many details of this process were not filled in until years after our course in 1961, these basic facts that Katz established about chemical synaptic transmission were what we learned listening to Potter and Furshpan's lectures. (Ironically, it turned out that a small but important minority of synapses *do* operate by the direct passage of electrical current, as Eccles had thought, and this additional mode of synaptic transmission had just been clearly established by Furshpan and Potter during their work as fellows in Katz's lab.)

As prospective physicians—and for me as a prospective psychiatrist—the biochemistry of neurotransmitters and their pharmacology was especially relevant because numerous drug effects depended on mimicking or inhibiting the action of the handful of transmitters that were then known. This instruction fell to Kravitz, the biochemist in Kuffler's group who was working on the chemical identification of the major transmitter in the mammalian brain that inhibits nerve cells from firing action potentials. The transmitter molecule (gamma-amino butyric acid) was obviously important in neural function, and Kravitz was later credited as its codiscoverer. Krayer, the head of the department and Kuffler's sponsor, taught the clinical pharmacology component of the course. With a phalanx of assistants, he set up elaborate demonstrations in anesthetized dogs to show us how various drugs affected the neural control of the cardiovascular system. These demonstrations were satirized mercilessly in the show that the second-year students put on, in which Krayer and his assistants, all in immaculate white coats, were depicted as German *geheimrats,* with thinly veiled Nazi overtones. None of us knew then that Krayer, who was not a Jew, had in the 1930s refused a German university chair vacated by a Jewish pharmacologist who had been dismissed under the Nazi laws prohibiting Jews from holding academic positions, or that based on his principles he had subsequently emigrated to England with the help of the Rockefeller Foundation.

Hubel and Wiesel rounded out this grounding in neuroscience by telling us how the brain actually used all this cellular and molecular machinery to accomplish things of interest to a physician. Unlike teaching us about the organization of nerve cells and the mechanisms of the action potential and synaptic transmission, conveying some idea of what the brain is actually doing was a difficult a task in 1961, and remains so today. Hubel and Wiesel dealt with this challenge by simply telling us about their work on the organization and function of the visual part of brain, which was just beginning to take off in a major way (see Chapter 7). It was not unusual for professors to cop out by telling us what they had been doing in their labs instead of going to the effort of putting together a broader and more useful introduction to some subject. But in this case, it was obvious that Hubel and Wiesel were trying to do something extraordinary, and none of us complained.

We managed to learn the organizational rudiments of the brain and the rest of the nervous system that we needed to know in complementary laboratory exercises in neuroanatomy (Figures 1.6 and 1.7).

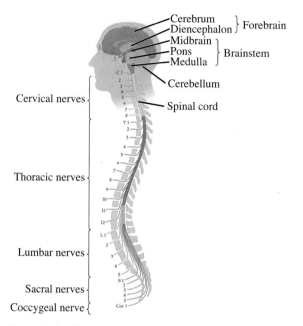

Figure 1.6   The major components of the human brain and the rest of the central nervous system, which is defined as the brain and spinal cord (After Purves, Brannon, et al., 2008)

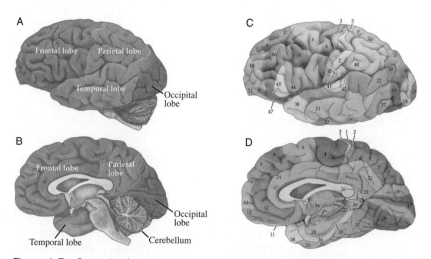

Figure 1.7   Some basic anatomical features of the human brain. The four lobes of the brain are seen in a lateral view of the left hemisphere (A) and a midline view of the right hemisphere after separating the two halves of the brain (B). (C) and (D) show the areas of the brain (each numbered) that German neurologist Korbinian Brodmann distinguished based on the microscopic studies of the cellular composition of the cerebral cortex. The views are the same as in (A) and (B), respectively. (After Purves, Augustine, et al., 2008)

By 1961, an enormous amount was known about the brain anatomy, thanks to the work of pioneers such as Cajál and Golgi, and of the neurologists and neuroanatomists such as Korbinian Brodmann who, in the early decades of the twentieth century, had devoted their careers to unraveling these details using increasingly sophisticated staining and microscopical methods (Figures 1.7C and D). The glut of information was made more tolerable by functional correlations that had been made between behavioral problems and deficits observed in patients with brain damage that was documented postmortem. These clinical-pathological correlations, which we heard about in the clinical lectures that were interspersed with the basic science Kuffler's group taught us, had been made routinely since the second half of the nineteenth century. Together with brain recording and stimulation during neurosurgery that had been carried out since the 1930s, clinical-pathological correlations showed in a general way what many regions of the brain are responsible for. Although Sherrington and others had confirmed and extended some aspects of the clinical evidence in experimental animals, prior to the work that Hubel and Wiesel were then beginning, neuroscientists could not say how individual neurons in the brain contributed to neural processing. Nor could they formulate realistic theories about the operation of an organ that, based on both its gross and cellular anatomy, seemed hopelessly complex.

For Hubel and Wiesel, as for many others before and since, the bellwether for understanding the brain was vision, and their goal was to figure out how the visual system works in terms of the neurons and neural circuits that the system comprises. To do this, they needed a method that would permit them to record from visual neurons in the brain of an experimental animal that, while anesthetized, was nonetheless responsive to visual stimuli. Hubel had devised a way to do this in 1957, using sharpened metal electrodes that would penetrate brain tissue and record the action potentials generated by a few nearby nerve cells. This method is different from the recording technique shown in Figures 1.3A and 1.4, in which the electrode penetrates a particular nerve cell and records the potential across its membrane (Figure 1.3B). An extracellular electrode simply monitors local electrical changes in brain tissue, and the results depend on how close the electrode is to an active nerve cell. In the hands of Hubel and Wiesel, this technique led to a series of discoveries about the mammalian visual system during the next 25 years that, as described later, came to dominate thinking about brain function in the latter half of the twentieth century.

Hubel, a Canadian who had been medically trained at McGill, had come to Hopkins as a neurology resident in 1954 and began to work with Kuffler two years later. Wiesel, a Swede, was medically trained at the Karolinska Institute in Stockholm and joined Kuffler as a fellow in ophthalmology the same year Hubel did. In joining Kuffler's lab as fellows in 1956, they benefited greatly from what Kuffler had already accomplished since coming to Hopkins in 1947 following his work with Eccles and Katz in Australia. Kuffler, always eclectic in the research he chose to pursue, had been working on the eye, in part because his appointment at Hopkins happened to be in the Department of Ophthalmology. In 1953, Kuffler had published a landmark paper in which he recorded extracellularly from neurons in the eye of anesthetized cats while stimulating the retina with spots of light, thus pioneering the general method that Hubel and Wiesel would soon apply to record from neurons in the rest of the cat visual system. Kuffler's key finding was that the responses of individual nerve cells in the retina defined the area of visual space that a neuron was sensitive to and the kinds of the stimuli that could activate it. (These response characteristics are referred to as a neuron's *receptive field properties*.) Hubel and Wiesel recognized that this same method could be used to study neuronal responses in the rest of the visual system. With Kuffler's encouragement, they began using this approach at Harvard in 1959 to examine the response properties of visual neurons in their own lab as junior faculty members. Based on what Kuffler had established in the retina, they were exploring the properties of visual neurons in progressively higher stations in the visual system of the cat, and this was the work that we students heard about as an introduction to brain function. Although most of it went over our heads, we all got the idea that the function of at least one part of the brain was being examined in a new and revealing way.

Our course on neuroscience ended in spring 1961, and we moved on to contend with the rest of the first-year curriculum. Despite the excellence of these extraordinary teachers, I soon forgot much of what I had learned. In the subsequent years, the information about the brain we were exposed to was more practical: In neuropathology, we learned about the manifestations and causes of neurological diseases. On rotations through the neurology, neurosurgical, and psychiatric services of various Boston hospitals, we learned about their clinical presentation and treatment. However, I understood that the group of neuroscientists Kuffler had brought together represented a remarkable collection of scientists working on problems that were importantly connected to my conception of what I might eventually do. But we all had to cope with the courses or rotations that were coming next, and I was still wedded to the idea of pursuing psychiatry. As it happened, this seemingly sensible plan began to disintegrate within a few months.

2

Neurobiology at Harvard

The scientists who introduced me to the biology of the nervous system in 1961 did not reappear in my life for another six years. For various reasons, I became increasingly uncertain about how to pursue my interest in the brain, if indeed I wanted to pursue at all it. In particular, my earlier conviction that psychiatry was a good choice began to wane as my knowledge of the field waxed during the next few years. The first disillusionment came in the summer after my first year in medical school, not long after the course that Kuffler and his recruits had given us. Although there was no rule against taking the summer break between first and second year as vacation, we were encouraged by the gestalt of med school—and a generous stipend—to spend the time working in one of the research labs at the school. Being zealots, many of us took advantage of the offer.

Our pharmacology course during the first year included several lectures on psychoactive drugs, and given my inclination toward psychiatry, I thought working with the professor who had presented this material made good sense. The professor, Dale Friend, was a respected clinical pharmacologist at the Peter Bent Brigham Hospital; I imagined I would learn a lot by working with him, and perhaps establish a useful liaison with someone in my intended specialty. Without making inquiries into what kind of a mentor he might be or who might have been better (I naively assumed that those who taught us were always the cream of the crop), I sought Friend out near the end of the term and he agreed to take me on as a summer student. However, shortly after I started working in his small lab in the hospital basement, he departed for an extended European vacation, and I was left to my own devices with only his lab technician as mentor. I came up with a cockamamie project that involved the measurement of norepinephrine (a neurotransmitter thought to be involved in depression) in the brains of rabbits given various doses of a popular antidepressant drug, and soldiered on. But the only significant result by the end of the summer was the demise of an unconscionable number of rabbits, and my sense that if this sort of work was typical of research in brain pharmacology, I wanted no part of it.

It was my first clinical rotation, however, that dispelled once and for all my earlier idea of pursuing psychiatry. Although we had some instruction in psychiatric diseases during the second year of medical school, our exposure to the psychiatry wards did not come until the third year. I was no stranger to psychiatric patients, having worked two summers in college as an orderly in a state mental hospital in Philadelphia, where I had grown up. It was interesting, if dispiriting, work that involved supervising very ill psychotic patients during the day, and making half-hourly bed checks and trying to stay awake on the night shift (which, as a summer employee, often fell to me). The male wards each had a padded room that was used to temporarily incarcerate patients who became unmanageable, and we orderlies were occasionally called on to wrestle a patient into a padded room and onto the floor so that a nurse could inject a massive dose of paraldehyde, a drug then in use that quickly put the patient into a stupor that lasted several hours. (The hospital also had a surgical suite for performing frontal lobotomies, and this approach to psychiatric treatment had only recently been abandoned.)

My third-year rotation in clinical psychiatry was at another state hospital, the Massachusetts Mental Health Center, housed in a ramshackle building a few blocks from the medical school. The range and severity of the disorders the patients suffered was not much different from what I had experienced as an orderly, but the treatment they were receiving was in some ways more discomforting than the padded room and paraldehyde method. Psychiatry at Harvard in 1963, particularly at the Mass Mental Health Center, was one of the last bastions of the Freudian analytical treatment of severely disturbed patients, and the preceptor during my rotation was a psychiatrist who believed strongly in its merits. Sitting through his analytically based interrogatories of psychotic patients eroded what little remaining faith I had that psychiatry was the field for me. The final blow came one Saturday afternoon when another student and I found a patient who had hung himself with his belt in one of the bathrooms. The usual resuscitative procedures were far too late to help the patient, but the psychiatry resident in charge clearly had less knowledge about how to proceed than we did, and dithered while we vainly undertook the steps for resuscitation that we had recently learned in other rotations. The message seemed clear: Psychiatrists were not real doctors, and being a "real doctor" loomed large for me, as it did for most of my classmates.

Anyone navigating the rigors of medical school gravitates toward one or more role models to provide inspiration. Having given up on psychiatry, I began to look elsewhere. Harvard abounded in such individuals, and one of the most charismatic in that era was Francis Moore, the chief of surgery at the Peter Bent Brigham Hospital, where I did my third-year rotation in surgery not long after my experience at the Mass Mental Health Center. Moore was then in his early fifties and already legendary. As an undergraduate at Harvard, he was president of both the Harvard Lampoon and the Hasty Pudding Club, and he had been appointed surgical chief at the Brigham at the age of 32. He was a member of the team that had performed the first successful organ transplant in 1954, and had written a widely respected book in 1959 called *The Metabolic Care of the Surgical Patient* that underscored his status as a physician–scientist of the first order. He had been motivated to write the book by the problems encountered with the burn patients he saw after the infamous Coconut Grove fire in 1942. Although the details in the book were largely incomprehensible to me when I read it as a student (and tried to use it later as a resident in surgery treating burn patients), holding retractors as a student assistant while Moore exercised his jocular authority and surgical skills turned me in a new direction. I decided then and there that I would train in general surgery. (Dramatic to the end, Moore committed suicide at the age of 88 by putting a gun in his mouth.)

After some further electives in surgery during my last year in medical school, I applied for a residency in general surgery at the Massachusetts General Hospital, considered the top program then, despite Moore's preeminence at the Brigham. I was accepted and started my surgical internship in summer 1964. It is hard to believe in today's atmosphere of managed care, oversight by insurance companies, and litigation over malpractice, that we "doctors in training" were largely unsupervised. Nevertheless, a strict hierarchy and rigid accountability existed within the cadre of surgical residents. In terms of workload and stress, that year was the most difficult of my life. The wards we were responsible for were always filled with patients who could not afford a private doctor, and, for better or for worse, they were entirely our responsibility ("better" was around-the-clock care by dedicated young doctors, "worse" was our lack of experience). Although a senior staff surgeon was nominally in charge, the attending, as he was called (no women were on the surgical staff), was in evidence only on morning rounds; he was rarely in the operating rooms, and then only on especially difficult cases at the invitation of the chief resident.

The chief resident I worked under during the first portion of my internship year was Judah Folkman. Much like Moore at the same age, Folkman was widely thought to be destined for great things, as turned out to be the case. At age 34, Folkman was named surgeon-in-chief at Boston's Children's Hospital, becoming, along with Moore, one of the youngest professors ever appointed at Harvard Medical School. He became justly famous in the early 1970s for pioneering a novel way of treating cancer by inhibiting blood vessel growth, and at 46 gave up this appointment as a professor of surgery to pursue basic research on angiogenesis full time. Folkman was perhaps the most impressive individual I have ever met. Physically slight, already balding, with a great nose inherited from his rabbi father, he dominated in any setting by intelligence, wit, and force of character. The residents universally looked up to him not simply because he was surgically skilled and supremely smart, but because he radiated a humorous self-confidence and integrity that made everyone under him better handle the daily strife and our inevitable mistakes. Folkman died prematurely in 2008, but inhibiting blood vessel growth remains a promising approach to some cancers and to other important diseases such as macular degeneration.

But working under Folkman that year had another effect on me. Competitive types continually measure themselves against the qualities and talents of their peers. As chief resident, Folkman was far advanced from an intern struggling to learn the rudiments of the trade. But he was only five years ahead of me, and I had to envision myself in his role in the near future. The comparison was discouraging. I didn't see him as necessarily smarter or feel that I could never reach his level of technical skill (although I had serious doubts on both counts). It was his obvious passion for the craft that dismayed me, a passion I had already begun to realize I lacked. I found reading the surgical journals as uninteresting as I had found reading Moore's book on postoperative care; and generating the zeal for operating that came naturally to Folkman was, for me, forced. The recognition that, as with psychiatry, a life in general surgery was probably not going to work for me came at 2 or 3 a.m. one night when Folkman was trying to make an apparatus to dialyze a patient dying from kidney failure using an old washing machine and other odds and ends he had collected from a hospital store room. Although the effort failed, I realized that I would not have made it, and that this disqualified me from trying to follow the footsteps of figures like Folkman and Moore.

And so I needed to invent another possible path to a professional future. The Vietnam War gave me some breathing space. By 1965, virtually all physicians in training were being drafted, and my notice arrived about halfway through my internship. Given my concern over the prospects of a career as a general surgeon, being drafted was not entirely unwelcome. There was another reason as well. Since the age of 12, I had suffered periods of depression. These bouts had tended to occur when my future direction seemed murky. A year of psychoanalysis during my last year of med school had not helped, although it contributed to my later decision that psychiatry, at least as it was then being practiced in Boston, was not something I wanted to pursue. Despite the modicum of professional training I had received by then, I thought of myself as neurotic instead of a person suffering from a clinical disorder, and that analysis might help; it didn't, and the diagnosis of depression did not seem to occur to the analyst I saw either. And so I ended my internship year not only without a clear plan, but clinically depressed. The bright side was that I would now have two years of enforced service to regain my balance and sort things out.

Despite the rapidly escalating war in Vietnam, the options for a young physician drafted in 1965 were surprisingly broad. I could join one of the armed services, which would have meant time in Vietnam and a domestic military base in some junior capacity; seek deferment for further specialty training and later service; apply for a research position at the National Institutes of Health; join the Indian Health Service; or apply to become a Peace Corps physician. This last option meant serving two years in the Public Health Service taking care of Peace Corps volunteers in one of many countries around the world. Given my uncertain frame of mind about what do to next, my lack of interest in research at that point, and my opposition (along with almost everyone else I knew) to the war, the Peace Corps seemed the best bet. And so after a few weeks of remedial training in tropical medicine at the Centers for Disease Control in Atlanta, I arrived in Venezuela in July 1965, without any idea of what my professional or personal future would be.

The Peace Corps turned out to be a good choice. The volunteers that I and another young doctor had to look after were an inspiring bunch who demonstrated all the good things about Americans and American democracy of that period. Because I spoke Spanish as the result of living in Mexico for four years as a kid and a further summer working in a rural Mexican clinic as a college student, I could travel easily and interact with local doctors. And Venezuela in the 1960s was a beautiful and relatively progressive country. From the medical perspective, the job was easy: I served as general practitioner to about 400 sometimes difficult-to-reach but generally interesting and healthy young adults. My life in South America was in every way a radical change, and for the first time since college, I had time to think about whatever I pleased instead of what I had to do next to meet the demands of medical training.

I read widely, including a lot of books on science that I had not been exposed to as an undergraduate majoring in philosophy, or as a medical student and intern with little extra time. One of the books I picked up while rummaging around the American Bookstore in Caracas was *The Machinery of the Brain*, by Dean Wooldridge, someone I had never heard of. He was an aeronautical engineer who, together with his Caltech classmate Simon Ramo, had left the aerospace group at Hughes Aircraft in 1953 to found the Ramo–Wooldridge Corporation, which later became the very successful defense-related company TRW. Having become wealthy, Wooldridge resigned from TRW in 1962 to pursue his passion for basic science, particularly biology. *The Machinery of the Brain*, published in 1963, was his first effort. By his own admission on the back cover of the book, Wooldridge's intention was simply to review the discoveries and ideas about the brain that he found interesting. Although I was passingly familiar with much of the content, his lucid synthesis of what was essentially the information I had learned about the brain as a med student got me thinking about issues that I had been keenly interested in when first exposed but had lost touch with as I morphed into a prospective surgeon. Unlike authors whose success in some unrelated field encourages them to pontificate about a new theory they imagine will explain the mysteries of the brain, Wooldridge's book was modest and provocative—and remains so as I look at it now—and it got me thinking about the nervous system once again.

Because I had two years to mull things over before I was scheduled to resume my post in general surgery at Mass General, I didn't need to rush to sort out my thoughts about a career that might combine an interest in the brain with my training as a physician. I even wrote a novel, on the premise that perhaps I had missed a literary calling (the few hundred pages of sophomoric prose I managed to crank out made clear that I had not). Finally, I concluded that the most logical course under the circumstances was neurosurgery. In addition to the internship year I had already completed, becoming a neurosurgeon required the further year of general surgery that awaited me on my return to Boston, so I was already well along this path. I thought that neurosurgery would combine my earlier and now reviving interest in the nervous system, which I had pushed aside as a result of my disillusion with psychiatry.

Soon after returning to Boston in summer 1967, I asked William Sweet, the head of neurosurgery at Mass General, if I might join his program when I completed my second year of general surgery. He agreed, and I was to start formal training in neurosurgery the next year. Despite the logic of this plan, within a few months of my return from the Peace Corps, I began to have doubts about what had seemed, in principle, an optimal marriage of interests and training up to that point. One of my rotations that year was as the general surgical resident assigned to neurosurgery, and I had looked forward to that stint. But disillusionment was not long in coming. Sweet, it turned out, was a caricature of the popular image of a brain surgeon and not the model I needed at that point to spur me on. He had a high-pitched voice and was something of a martinet, making him an easy target of mockery by the residents, one of whom was especially good at mimicking his prissy persona. Sweet was no Moore or Folkman, and his colleagues on the neurosurgery faculty at the time were not much more inspiring. When I experienced it first hand, their daily work was not all that interesting. In contrast to my abstract enthusiasm for exploring the brain in this way, the actual operations were long and tedious, and the outcomes (especially in patients with brain malignancies) were all too often a foregone and unhappy conclusion. Some procedures—such as evacuating a hematoma, clipping a ruptured blood vessel, or removing a benign tumor—were curative but didn't do much to stoke my interest in understanding the brain in some significant way. Even the psychosurgery then being carried on at Mass General was unappealing, because of both its oddball practitioners and the flimsy scientific grounds on which such work was being justified.

As I worried increasingly about the prospect of neurosurgery, my thoughts kept turning back to the neurobiologists Kuffler had brought to Harvard in the late 1950s and the impression they had made on me as a first-year student. In winter 1967, I found myself back at Harvard Medical School in the office of David Potter, who had taught us about action potentials and synaptic transmission six years earlier. I remembered Potter as the most approachable of the group and sought him out for advice about whether research in neuroscience might be a reasonable option. He listened with apparent enthusiasm as I summarized my concerns with neurosurgery and my interest in perhaps giving research in neuroscience a try. On the face of it, my arguments were feeble: The sum total of my research experience was the disastrous summer spent in Friend's pharmacology lab, and my desire to try research had been reached largely by excluding other options. Despite the manifest weakness of my case, Potter agreed to think about my situation, and we agreed to meet again. In the meantime, I asked Sweet if I might take my first year in the neurosurgery program as a research fellow, and, primarily because of scheduling issues with other residents in the program, he agreed.

When I returned to Potter's office a couple of weeks later, he had indeed given the matter some thought. He considered my intention to try research plausible enough and suggested that I contact John Nicholls (Figure 2.1) to ask if he could take me on as a fellow. Kuffler had just recruited Nicholls to Harvard but he was still working at Yale as an assistant professor. (Kuffler's original group was then in the process of expanding and to form a full-fledged department.) I was disappointed because I had no idea who Nicholls was and had hoped to work with Potter or perhaps even Hubel and Wiesel, whose stars were rising even when I was a student. I was even more dismayed when Potter told me that Nicholls was working on the nervous system of the medicinal leech. I had no idea what any of the neurobiology faculty was doing then or why, but it was difficult to imagine how the leech could be pertinent to my ill-formed ambition to become a neuroscientist who might ultimately say something important or at least relevant to the human brain.

Figure 2.1   John Nicholls circa 1975. (Courtesy of Jack McMahan)

In fact, Potter's suggestion was a good one. I was somewhat reassured when Potter told me that Nicholls had been a graduate student with Bernard Katz in the late 1950s after he had completed his medical training in London, that he had been a fellow in Kuffler's lab thereafter, and that his work on the leech was widely regarded as an outstanding example of what was then a new approach to understanding neural function—studying the nervous systems of relatively simple invertebrates. Potter went on to say that because Nicholls would be starting up a new lab at Harvard, he would probably welcome a fellow, even one whose experience in neuroscience was nil. As a result of this conversation I wrote to Nicholls, who invited me to visit him at Yale to meet and talk things over.

And so on a bleak Saturday in February 1968, Shannon Ravenel, whom I was going to marry later that spring, and I drove down to New Haven. I had met Shannon while still in med school, and we had an on-again, off-again relationship that had, happily for me, been on-again since I returned from Venezuela. Yale Medical School was unimpressive (I had never actually been there, even though it was only a few blocks from the residential college where I had lived as an undergraduate). Nicholls's lab was equally nondescript, and it was quickly evident that he had a complex personality that might not be such a good fit with my own. He asked us to dinner at his modest apartment, where his two unruly children cavorted about in their underpants, and where what appeared to be a dysfunctional relationship with his wife was palpable. Driving back to Boston that night, Shannon pointedly asked me if I really wanted to make this change in light of all the evidence that I would be sailing into unknown and possibly stormy waters. Even though my confidence in answering was minimal, several reasons argued for seeing it through: Potter's word that working with Nicholls was a good bet, the lack of obvious options if I wanted to try research, and a determination on my part to do something that might reignite a passion for understanding the brain. And I could always go back to the neurosurgery program at Mass General if the research year failed; that was Sweet's expectation, even if I had my doubts.

Another factor in going forward was Denis Baylor, a research fellow whom I had briefly met in Nicholls's lab. Baylor had gone to medical school at Yale; like me, had decided to try his hand at neuroscience research; and had ended up spending three years with Nicholls as his sole postdoctoral collaborator (not an unusual arrangement in those days). When we were alone, he told me in no uncertain terms that Nicholls had been a terrific mentor and friend. Baylor, who later went on to work with Alan Hodgkin at Cambridge and had a stellar career investigating the properties of photoreceptors in the retina, was clearly sensible, and his encouragement counted for a lot. So a few days after getting back to Boston, I called Nicholls to say that I was willing if he was, and he agreed that I could take Baylor's place when he arrived at Harvard later that summer. Shannon and I married in May, I finished my second year of residency in June, and after spending that summer in Vietnam under the auspices of a Boston-based antiwar group selecting war-injured children to come to the United States for treatment, at age 30 I began life as a neuroscientist.

Although the Department of Neurobiology at Harvard was the best place I could possibly have tested the merits of this new direction, the transition was not easy. For the previous four years, I had been a practicing doctor, and whether in Boston, Venezuela, or Vietnam, I had all the responsibilities and respect that being a physician entails. Suddenly I was a superannuated student on the bottom rung of the ladder; even the two beginning graduate students in the new department knew more science than I did, and they seemed a lot smarter to boot. The stress was of a very different kind than I had experienced during the year I had worked as a surgical intern, but the first year I spent in Nicholls's lab was in some ways nearly as trying, with one fundamental difference: Despite my ignorance and well-justified sense of inferiority, I finally loved what I was doing. For the first time in years, I worked hard not because I had to, but because I wanted to.

The approaches to the brain and neural function that Kuffler and his young faculty were spearheading when I was a student in 1961 had flowered by the time I returned as a fellow in 1968. During the first half the twentieth century, the major goals in neuroscience had been reasonably clear: to determine how action potentials work, and to understand how information is conveyed from one nerve cell to another at synapses. Because it was obvious that all brain functions depend on these fundamental processes and their cellular and molecular underpinnings, it didn't make sense to grapple with more esoteric issues until these central challenges had been met (although, of course, other research was going concurrently, primarily with the goal of better understanding the organization of the brain). By the early 1960s, however, Hodgkin and Huxley had deciphered the mechanism of the action potential, and Katz and his collaborators had convincingly demonstrated the basic mechanism underlying chemical transmission at synapses. The question that confronted the next generation of neuroscientists—Nicholls and the rest of the faculty—was what to do next.

In determining the possible directions of neuroscience, Kuffler and his eclectic style of research was a powerful force. It helped that he had collaborated with Hodgkin, Huxley, and Katz in the 1940s, and that he was, in his own different way, an intellectual giant. One would never have guessed this, however, from his inability to give a coherent lecture, his penchant for bad puns, and his democratic approach to everyone in the department, whom he insisted call him "Steve" (Figure 2.2). His only affectation, a minor one to be sure, was always wearing a white lab coat.

Figure 2.2   Steve Kuffler in the department lunchroom, circa 1970. (Courtesy of Jack McMahan)

Kuffler's study of the retinal cell responses at Hopkins, which was the impetus for Hubel and Wiesel's work on vision, had already supplied one general answer to the question of what to do next. By the time I arrived back at Harvard as a fellow, Hubel and Wiesel were already well on the way to the dominant position in brain physiology that they would hold, for very good reasons, for the next several decades. Everyone working in neurobiology in the late 1960s assumed that it was just a matter of time until Hubel and Wiesel, and the cadre of followers they were beginning to spawn would provide a deep understanding of vision and perception.

Another aspect of Kuffler's work had stimulated a different direction that seemed equally promising, and this had determined the work that Nicholls was doing when I joined his lab. Despite the growing success of the experiments that Hubel and Wiesel (and others) were carrying out in the brains of cats (and later in monkeys), a concern in the 1960s was that the brain might be too complex to readily give up the secrets embedded in the details of its circuitry. Kuffler always had a knack for picking relatively simple systems such as the retina, the neuromuscular junction, or the sensory receptors found in muscles as a way to unravel some problem in neuroscience and extract answers of general significance. His approach was not unique—Hodgkin and Huxley had used the squid giant axon to understand action potentials for similar reasons, and Katz had focused on the frog neuromuscular junction as a model system in which to understand synaptic transmission. But Kuffler was especially insistent about the value of using a variety of simple preparations to advance the cause, and one of these preparations was the nervous system of the medicinal leech (Figure 2.3). Following his graduate work with Katz, Nicholls had joined Kuffler's lab as a fellow in 1962, and they had worked together to understand the function of the other major class of cells in the nervous systems of all animals, the non-neuronal cells called *glia*. Glial cells outnumber neurons in the brain by three or four to one, and their role was largely a mystery; therefore, it made sense to look at what they were doing and their relation, if any, to the functions of nerve cells. For various technical reasons, Kuffler had decided that the leech was the best animal in which to carry out such work. By the time Nicholls left Harvard to join the faculty at Yale, the work on glia had largely finished, but he continued using the leech as a simple system in which to explore neuronal circuitry in relation to behavior.

The opinion held by Nicholls and many other neuroscientists when I joined his lab in the fall of 1968 (the only other lab member at that point was Ann Stuart, a graduate student earning her Ph.D.) was that a logical next step in moving beyond the established understanding of neural signaling would be to focus on invertebrate nervous systems as models for fathoming the basic principles of neural organization and function. The nervous systems of the leech and some other invertebrates were attractive in terms of neuronal numbers (hundreds or thousands compared to the 100 billion or so nerve cells in the human brain), large enough to enable the identification of the same nerve cell from animal to animal using only a low-power microscope, and easily recorded from with intracellular microelectrodes (see Chapter 1). The prospect of relating the function of identified

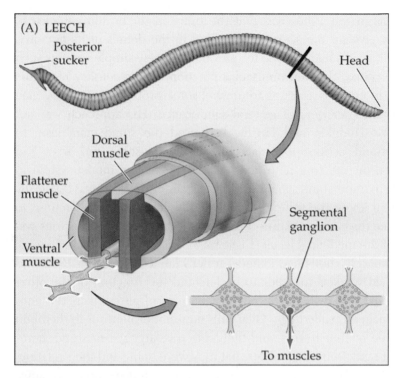

Figure 2.3   The medicinal leech and its central nerve cord (the chain of ganglia shown at the bottom right), the preparation that Kuffler and Nicholls used to study the role of glial cells in the nervous system. (After Purves, Augustine, et al., 2008)

nerve cells to some bit of behavior seemed both attractive and straightforward. Nicholls, Baylor, and Stuart had been plugging away with this goal when I visited Nicholls at Yale, and I threw myself into it as a novice neuroscientist at Harvard.

Absorbing the rationale for exploring the nervous system in this way, and learning the required methods and their scientific basis as a beginner with no background in electronics, mathematics, or anything else that was particularly relevant, led to many low moments. But the work was conceptually and technically fascinating, and I was buoyed by the fact that a lot of obviously smart people thought this was a good way to explore how vastly more complex brains do their job. In addition to Nicholls, the members of the department then operating under the assumption that a route to further success in neuroscience lay in mining answers that could be provided by simple preparations were Ed Kravitz, working on neurotransmitters in lobster ganglia (Figure 2.4A); Potter and Furshpan working on the synaptic function of small numbers of isolated nerve cells in tissue culture; Zach Hall working on the molecular components of synapses; and Kuffler and his group working on the nervous systems of leeches, mudpuppies, crayfish, and frogs with a variety of related interests and aims.

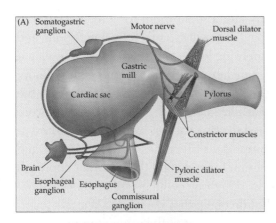

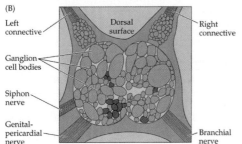

Figure 2.4   Some other "simple" central nervous systems being studied in the late 1960s under the assumption that invertebrate preparations would be a good way of understanding the operating principles of more complex nervous systems. A) The "brain" and stomatogastric ganglion of the lobster. B) The abdominal ganglion of the sea slug. The perceived advantage of these preparations compared to mammalian brains was the small number of cells, many of which could be individually identified and studied by intracellular recording. (After Purves, Augustine, et al., 2008)

The only people in the department actually working on mammalian brains were Hubel and Wiesel. Although they had begun as Kuffler's protégés in the late 1950s, by 1968, they had established a largely separate unit within the department, which they ran according to their own lights. Hubel and Wiesel were attentive to and interested in what others in the department were up to, but they seemed to see their own goals as different (and, one sensed, more important) than those represented by the work that the rest of Kuffler's faculty was doing. The lab assigned to Nicholls when he arrived was in the same limb of the U-shaped department that housed Hubel and Wiesel's separate empire (in reality, just a few contiguous lab rooms, but with space for their own secretary and a small room for a technician who did photography and histology). They also had a little lunchroom, and Nicholls, Stuart, and I routinely ate lunch with Hubel and Wiesel and the four or five people in their lab. So I came to know more about them and their work than the others in the department. Although the respect everyone had for Kuffler kept a lid on things, strains were apparent. The tension seemed to arise not just from personal idiosyncrasies, but from the philosophical differences between Hubel and Wiesel's direct approach to brain function and the different approach of the majority who were working out cellular circuits and mechanisms in simpler systems. It didn't help that Hubel and Wiesel's scientific success was rapidly outstripping that of the other members of the department.

The enthusiasm for simple invertebrate systems as a way to understand aspects of neural function eventually paid off, but not in the ways that were envisioned in the 1960s. The dominant idea then—that relating the properties of individual neurons to neural circuits and behavior would yield insights into the functional principles of complex brains—died a slow death during the following 20 years. There were several reasons, but a primary cause was the rapid rise of molecular biology that followed the discovery of the structure of DNA in the early 1950s. It was apparent by the mid-1970s that the tools of molecular biology provided a powerful way to study neural function. James Watson, in his typically provocative way, proclaimed publicly that molecular biology would solve the remaining problems in neuroscience within 25 years. Although Watson's claim was nonsense, the entry into neurobiology in the late 1960s and 1970s of superb molecular biologists such as Sydney Brenner, Seymour Benzer, Marshall Nirenberg, Gunther Stent, Julius Adler, Cyrus Levinthal and others (Francis Crick ultimately joined the crowd) rapidly changed the scene with respect to simple nervous systems. With surprising speed, the neurogenetics and behavior of rapidly reproducing invertebrates, such as the roundworm *Cenorhabitis elegans* and the fruitfly *Drosophila melanogaster,* supplanted leeches, lobsters, and sea slugs as the invertebrates of choice, with molecular genetics considered the best way to ultimately understand their nervous systems. Even the success of Eric Kandel, who started working on the sea slug nervous system (Figure 2.4B) in the 1960s for the reasons then in fashion, was ultimately based on how the modulation of the genes by experience can lead to synaptic changes that encode information. Kandel, who won a Nobel Prize in 2000 for this work, was the most energetic cheerleader for the simple central nervous system approach when I arrived in Nicholls's lab. The first lunchtime seminar I presented in the Department of Neurobiology in 1968 was a critique of one of Kandel's early papers. Nicholls saw Kandel as a competitor whose work did not meet the standards of scientific rigor that he had absorbed as a student of Katz and Kuffler. Even though I had been at it only a few months, Nicholls's bias had already rubbed off on me. Nonetheless, I quickly gathered that there was no consensus among my new colleagues about how to go about understanding the brain and the rest of the nervous system.

As Baylor had promised, Nicholls was indeed a good mentor. After two years of working hard to understand how sensory and motor neurons in the leech's nervous system were related, I had written two papers of very modest interest that were published in the *Journal of Physiology*, the journal of record in those days (an output of one detailed paper a year was typical then, although it wouldn't get anyone a job today). But despite my appreciation of his teaching, Nicholls and I did not get along particularly well. In addition to what we were doing on a daily basis, I wanted to discuss the broader issues that had always interested me about brains and their functions, and Nicholls didn't have much stomach for that sort of thing. He told me that as a graduate student, he had been terrified of Katz and the prospect of his failing in his eyes; perhaps as a result, he seemed unwilling to think in grander terms that Katz would presumably have thought overblown or silly. Our personalities and perspectives were very different in other ways as well. One day Hubel and Wiesel called us into their lab to show us the response properties of a particularly interesting neuron that they were recording from in the visual cortex of one of their cats. There was not much doubt by then that what Hubel and Wiesel were doing would influence thinking about the brain for a long time to come, and some humility in the face of this presumption was certainly warranted. But I was taken aback when Nicholls remarked after we went back to the lab that some of us in science inevitably had to be the drones and not the queen bee. Although empirically true, to hold this view while still in one's thirties (Nicholls was then 36) seemed to me a bad idea.

By mutual consent, Nicholls and I agreed that I would spend the third year of my fellowship in Kuffler's lab (I had already told Sweet that I would not be returning to the neurosurgical program). During my last months with Nicholls, I had been working with Kuffler's senior collaborator at the time, Jack McMahan. We had been looking at the anatomy of leech neurons revealed by the injection of a fluorescent dye that Kravitz had developed as a sort of sideline. This was an exciting new method that for the first time enabled neuroscientists to see all the axonal and dendritic branches of an identified nerve cell whose electrical properties had been studied by recording through the electrode that injected the dye. Jack, who is a terrific neuroanatomist, was about my age, although he was years ahead of me as a neuroscientist and about to be named to the faculty. We got along famously and had a wonderful time that year, during which he taught me how to do electron microscopy. Unlike Nicholls, who was always reticent to simply natter on about neuroscience, Jack and I argued incessantly about what might be done and how to do it.

During that year, Katz visited Kuffler's lab, as he regularly did. The friendship and mutual respect between the two formed in Australia had resulted in a small but steady flow of young neuroscientists between Boston and London. It was clear that working in Katz's orbit would be a fine next step for me. Although I had learned a lot of neuroscience in my three years as a fellow, I had started from scratch. It seemed foolhardy not to seek more training before going off on my own, and no one seemed to disagree. Moreover, I had decided that working on invertebrate nervous systems was not the road to the future I wanted to follow; working as a fellow on a different topic would therefore be important. And so I asked Nicholls and Kuffler if they would support my case to Katz, which they did. Harvard offered a generous traveling fellowship that supported two years of study abroad, and in the summer of 1971, Shannon and I set off for London with our one-year-old daughter. What I would pursue in Katz's small Department of Biophysics at University College had not been specified, but I was at least sure by then that I had found the right profession.

3

Biophysics at University College

Although Bernard Katz's impact on the course of neuroscience was as at least as great as Kuffler's, the style of the two men was very different, and their influence was felt in different ways. Whereas Kuffler's modus operandi was quintessentially eclectic—he would work on a project with a collaborator or two for a few years and then move on to an entirely different problem—Katz was a scientific bulldog. He had seized on the fundamental problem of chemical synaptic transmission in the late 1940s and never let it go. And whereas Kuffler was, superficially at least, an extroverted democrat, Katz was reserved and, to some degree, an autocrat.

As a result of these personal contrasts, as well as the cultural distinctions between the way science was practiced then in the United States and the United Kingdom, Katz's Department of Biophysics at University College London was very different from the Department of Neurobiology at Harvard. Although World War II had ended 25 years before I arrived in 1971, the mentality of rationing and general tight-fistedness remained. The labs were on the upper floors of one of the old buildings on Gower Street that ran the length of a long London block and housed most of the basic science departments. The rooms of the five faculty members in the department were comfortable but modest, and the surfeit of furnishings, equipment, and supplies that I had been used to in Boston was nowhere in evidence. Even Katz's lab was outfitted with equipment that would have been consigned to storage at Harvard, and his small office contained the same simple furniture that must have been used by A. V. Hill when he was director in the 1930s. The only notable feature was the bronze bust of Hill that stared down from its place of honor on a bookshelf.

Hill had been "A. V." to Katz, and Katz was now deferentially addressed as "B. K." by the faculty, and as "Prof" by the staff.

The departmental infrastructure was run with an iron hand by a tenured technician named Audrey Paintin. She was a fireplug of a woman who demanded an extraordinary degree of diplomacy on the part of her petitioners to obtain even the most basic supplies. The work of one of the fellows who arrived about the same time I did was blocked for weeks because his robust American style did not suit Paintin, and even Katz seemed to tread lightly when dealing with her. The cost of everything was carefully considered. In addition to Paintin's frugality, Katz's secretary told me in no uncertain terms a few days after my arrival that notes to her or anyone else would be better written on back of a piece of paper that had been used for another purpose than on a fresh sheet. Among other things, this proved that superb science had been and could be done in very modest circumstances.

The students and fellows in Katz's domain (about a half-dozen of us in total) were in labs along a short corridor on the floor that included Katz's lab at one end, as well as Paintin's supply room and a machine shop. For most of the 1960s, Katz had collaborated with Ricardo Miledi, an extraordinarily talented experimentalist who was Katz's executive lieutenant and the most prominent of the other faculty members. Miledi had a smaller lab adjacent to Katz's where he pursued his own projects with a couple of fellows, in addition to his ongoing work with Katz. The other faculty members were on the floor above and included Paul Fatt, a brilliant but eccentric physiologist who had collaborated with Katz in the early 1950s in discovering the quantal nature of chemical synaptic transmission (the fact that neurotransmitters are packaged in synaptic vesicles released from axon terminals by the arrival of an action potential; see Figure 1.4B); Sally Page, an electron microscopist; and Rolf Niedergerke, another very good biophysicist and a disciple of Andrew Huxley who was working on the properties of heart muscle. The faculty in British universities operated far more independently than their counterparts in the more casual and collegial atmosphere of U.S. departments, typically behind closed doors. This unfortunate tradition presumably stemmed from the tutorial system of Dons that had been practiced for centuries at Oxford and Cambridge. Given this academic style, the faculty members upstairs were rarely in evidence on the floor below, even at teatime.

Although I expected to work on a project that would explicitly tap into the expertise and interests of Katz, Miledi and the others in this new environment, I had no idea what my options would be when I arrived in London. The year I had just spent working with Jack McMahan in Kuffler's lab had been extraordinarily valuable. Among other things, it had introduced me to vertebrate autonomic ganglia, the small collections of accessible and quite beautiful nerve cells that Kuffler and his postdocs Mike Dennis and John Harris were working on in the frog heart, and that Jack and I had studied in the ganglia of the mudpuppy heart. But my work with McMahan had not presented a problem that seemed especially worth pursuing. Having already soured on the nervous systems of simple invertebrates such as the leech, I needed to determine what general direction in Katz's department would make sense and give me a starting point for my own research in the academic job I would have to secure when my fellowship ended.

The first day I came in to work in the department after getting settled in our flat in Hampstead, Katz invited me into his office to discuss what I might do. Katz, then 60, was austere but certainly not the terrifying figure John Nicholls had described (Figure 3.1); I never knew whether this difference reflected a softening of Katz's style with age and success, or Nicholls's neuroses. Katz listened to my ill-formed ideas about issues that I might want to consider, and said that I should take my time in deciding on a particular course. He suggested that I also discuss the prospects with Miledi, who rode herd on what all the fellows were doing, and mentioned that another postdoc, Bert Sakmann, happened to be at loose ends and was also thinking about what to do next. Sakmann had just finished up a year-long project with Bill Betz, another fellow who was about to return to a job in the States, and Katz thought that it might make sense for us to work together on something of mutual interest. I met Bert later that day and we chatted about the possibilities.

Bert (Figure 3.2) had been medically trained in Tübingen and later in Munich, where he said he had gone in pursuit of the fellow medical student he eventually married. In the course of his medical education in Munich, Bert had spent three years carrying out research on the visual system with Otto Creutzfeldt. Similar to many of us brought up scientifically in that era, Bert thought that working directly on the visual system or some other part of the brain was a

Figure 3.1   Bernard Katz at the Missouri Botanical Garden on a visit to St. Louis circa 1980. (Courtesy of David Johnson)

rather daunting prospect. With Creutzfeldt's help, he had sought out further training with Katz to pursue a future working at the seemingly more tractable level of neurons and their synaptic interactions. We hit it off well because of our similar backgrounds, shared fascination with all aspects of neuroscience, and corresponding opinions about the odd cast of characters and their relationships in the Department of Biophysics. Although Bert was four years younger, we were both recently married, ambitious, and faced the need to land academic jobs when we finished our fellowships in two years. We wanted to do something significant that would get our careers off and running, but our initial conversation made clear that neither one of us had a very good idea about what this might be.

In 1971, Katz and Miledi were nearing the end of their extraordinarily productive collaboration on the mechanism of synaptic transmission. Katz had been awarded the Nobel Prize in Physiology or Medicine the year before for his fundamental series of discoveries about synaptic transmission, and this line of research using electrophysiology and complementary electron microscopical methods

Figure 3.2   Bert Sakmann at a much older age, but much as he appeared in the early 1970s. (Courtesy of Bert Sakmann; photo by Sven Erik Dahl)

seemed almost complete after more than 20 years. Katz and Miledi were starting a new phase of investigation at the molecular level. Their immediate aim was to understand the molecular events underlying the action of neurotransmitters by looking at what was then called "synaptic noise." The noise in question was tiny, random fluctuations of membrane voltage that they thought might represent the effects of transmitter molecules opening individual ion channels at the synapse. Identifying neurotransmitter action at this level would be another major advance and would open a new chapter in understanding synaptic mechanisms. At the same time, Miledi and Lincoln Potter, a young biochemist who was David Potter's brother, were undertaking an ambitious effort to isolate and identify the receptor molecule for acetylcholine, the neurotransmitter at the neuromuscular junction. These projects, which were already highly competitive and fraught with controversy, seemed to be far beyond our skills or interests. (Ironically, solving the noise problem, at which Katz and Miledi ultimately failed, would make Sakmann famous within a decade.)

The issue that captured our attention, and that of many other neuroscientists at the time, was how the neural activity generated by everyday experience affects synaptic interactions and neuronal connectivity. It had long been apparent that understanding how the effects of experience are encoded in the nervous system presented another major challenge in neuroscience. Successfully addressing this issue would explain the way we and other animals learn. And unlike the mechanisms of neural signaling, this problem was far from being solved. Humans and many other species are obviously changed by what happens to us in life, and the lessons learned are an important determinant of evolutionary success. For all but the simplest organisms, modification of the nervous system through learning contributes importantly to the efficacy of behavior, and ultimately to the likelihood of reproducing. Because the currency of experience in neural terms is the action potentials that stimuli generate through the agency of sense organs, it had been widely assumed for decades that learning involves activity-dependent changes at synapses. These changes—referred to more generally as synaptic plasticity—would therefore encode new information by actually or effectively altering neural connectivity. Transient changes that reflected alterations in the efficacy of synaptic transmission would presumably explain short-term learning and memory (such as a telephone number read from the phonebook and remembered for only as long as needed to dial it). More permanent anatomical changes in synaptic connectivity seemed likely to be the basis for long-term memories that could last for years (such as remembering your telephone number in childhood).

While I was still at Harvard, a handful of us had organized a series of evening meetings to identify pathways in neuroscience that seemed promising avenues we might follow when eventually we had independent labs. A recurrent theme in these inconclusive conversations was figuring out how activity changed neural circuits. Pursuing some aspect of this challenge seemed a worthy goal to both Bert and me. This had been one of the major purposes in much of the work on the leech and other simple central nervous systems (relating changes in the connectivity of single identified neurons to changes in observed behavior), and thinking along these lines was not much of a stretch. Katz had pioneered studies of this general problem in the 1950s by asking how the release of neurotransmitter at the neuromuscular junction, the model synapse illustrated in Figure 1.4A, was affected by prior activity. The result showed that the efficacy of neurotransmission could be either increased or decreased, depending on the nature of the preceding activity: Small amounts of activity facilitated transmitter release, and large amounts depressed it, a phenomenon that was followed by a later rebound increase that lasted minutes or longer.

This general perspective had already motivated a lot of related work in other labs in the 1960s, and one of these was the lab of Per Andersen in Oslo. Andersen, along with Kuffler and Katz, was a trainee of John Eccles, albeit many years later (the roster of important people Eccles trained during several decades is a remarkable testament to his impact on the field, whatever one might think of his sometimes odd and stridently expressed views). A student of Andersen's named Terje Lømo had discovered a particularly long-lasting form of potentiation in the brains of rabbits in 1966, a topic he pursued with Tim Bliss, another fellow who had arrived in Andersen's lab in Oslo in 1968. The phenomenon that Lømo and Bliss described was in the hippocampus, a region of the brain known to be involved in a particular form of human memory, and was rightly taken to be especially important. Long-term potentiation in the hippocampus eventually spawned hundreds of research papers, and its role in memory continues to be a topic of intense interest (and ongoing controversy) today.

Lømo was a fellow in Katz's department at the time Sakmann and I were considering what we might do, but he was about to leave to work further with Bliss at Mill Hill in north London, where they pursued hippocampal potentiation and firmly established the importance of this line of investigation. However, Lømo had worked on a different project at University College with Jean Rosenthal, another fellow who had just left. Together they had shown that prolonged stimulation of a muscle changed the sensitivity of the muscle cells to the neurotransmitter (acetylcholine) that normally activates these cells by release at the neuromuscular junction. This effect was also of obvious interest because it suggested another way of exploring how activity could change the behavior of excitable cells, and therefore how the nervous system might encode information derived from experience. After some further discussion, Bert and I decided that following up on what Lømo and Rosenthal had done would be a fine way to better understand how activity could alter the properties of nerve and muscle cells and, in principle, store information.

After hashing it over, we discussed this general goal with Miledi (Figure 3.3). Miledi's work in the 1960s done independently of Katz had shown that when the innervation of a muscle was removed by cutting the nerve to it, the sensitivity of the muscle fibers to neurotransmitter acetylcholine, which was normally limited to the immediate region of the synapse, spread over the whole muscle fiber surface. This observation had set the stage for Lømo and Rosenthal's demonstration that directly stimulating the muscle could reverse this "supersensitivity." Not surprisingly, Miledi was keen when Sakmann and I indicated that further work along these lines piqued our interest.

Figure 3.3   Ricardo Miledi with Katz by their experimental rig in the early 1960s. Although room temperatures in the department were always on the cool side, they had purposely turned off the heat for the experiments they were doing that day. (Courtesy of Ricardo Miledi)

Our idea for a way to attack this issue was based on an odd fact that neurologists had known and used as a diagnostic tool for decades. When muscle fibers are no longer innervated (a common enough occurrence in human injuries or diseases such as polio or amyotrophic lateral sclerosis), the muscle fibers begin generating action potentials on their own, a phenomenon called "fibrillation." The origin and consequences of this spontaneous activity raised some further ways to explore the control of nerve and muscle cell membrane properties by activity (the occurrence of action potentials), and these could be studied in muscles taken out of an animal such as a rat and kept alive for a week or more in a Petri dish. In these circumstances, we could directly monitor the levels of spontaneous activity in individual muscle cells with a recording electrode, experimentally alter the levels of activity by electrical stimulation or blockage with a drug, and test the sensitivity of the fibers to neurotransmitter. Katz and Miledi agreed that this would be a sensible project. So for the next two years, we happily set about exploring these issues.

Because neither we nor anyone else in the department knew exactly how to go about this, we fiddled with various chambers, muscles, methods of stimulation, recording electrodes, and culture conditions until we got things to work. Eventually, we could record for several days from single muscle fibers and watch their activity wax and wane as spontaneous action potentials or their absence in the fibrillating fibers altered their membrane properties and, consequently, their sensitivity to neurotransmitter that we applied. We could also stimulate the muscle artificially, showing that denervated fibers that were kept active never started to fibrillate, and could be made to stop fibrillating if this spontaneous activity had been allowed to start. Although the results were another modest contribution (a couple of good but rarely cited papers in the *Journal of Physiology*), we had a fine time being on our own and doing what we thought was interesting.

Katz would drop by every few days to see what was going on, but rarely offered detailed advice and was usually more interested in chatting about neuroscience generally, politics, and the grand scheme of things (which we were delighted to do). He was then Secretary of the Royal Society and thoughtfully arranged for us to go to occasional evening meetings, pompous but interesting events where members presented demonstrations in much the same way they had since the Society's founding in the seventeenth century. (We had to wear rented tuxedos to attend.) Miledi also dropped by our lab from time to time, usually offering specific advice about complicated variants of our experiments that he thought we should try, advice that I don't think we ever followed. His oversight was well meaning, but I soon came to appreciate the extraordinary clarity and focus of Katz's thinking compared to the rest of his faculty. The most important feedback we got was from the other fellows who were there at the time, particularly Nick Spitzer, Mike Dennis, and John Heuser, who were all transplants from Harvard like me.

Katz gave the two manuscripts that Bert and I wrote up at the end of our time in the department his seal of approval and suggested we show the papers to Huxley, who was in the Department of Physiology a few floors and corridors away in the warren of University College buildings. He had been studying muscle contraction since the 1950s in work that was as impressive in its own way as what he had done with Alan Hodgkin on the action potential in the 1940s. Because our work concerned muscle cells, Katz thought Huxley would be interested, and that he might have some useful criticisms. Huxley deemed the papers more or less fine, but he chastised us for having blacked out some fuzziness around the oscilloscope traces with a marker—an innocuous bit of pre-Photoshop-era improvement to our figures that, for a purist like Huxley, was a cardinal sin.

In the end, our apprenticeships in neuroscience at University College had not provided either Bert or me with a compelling problem to follow up. Although the work we had done was perfectly good, it didn't present us a clear path to the future, in the way Lømo's work on potentiation in the hippocampus led him toward a specific goal and early notoriety. There is no formula for figuring out what to do in science, and everyone who eventually finds a good problem does so in a different way, if they find one at all. Bert left London in 1973 a month or so before I did to begin an assistant professorship at the University of Goettingen. Within the year, he began a collaboration with Erwin Neher that eventually led to the Nobel Prize for Physiology or Medicine in 1991. The award was for making possible a further key step in understanding the basis of neural signaling that Hodgkin, Huxley, Katz, and Kuffler had done so much to advance in the preceding 30 years. Although it had long been clear that both the action potential and synaptic transmission depended on changes in the movement of specific ions though channels in nerve cell and muscle fiber membranes, the details of how this actually occurred had remained uncertain. The major obstacle was the absence of a way to measure what everyone presumed was the opening and closing of these channels in the cell membranes of nerve and muscle cells, activated by either voltage changes across the membrane in the case of the action potential, or by the action of a neurotransmitter at synapses. This was the issue that Katz and Miledi had been working on in their studies of synaptic noise that were going on down the hall. Sakmann and Neher solved this problem by the ingenious trick of pulling a small patch of membrane into the mouth of a highly polished electrode, which enabled them to record the tiny electrical events associated with the opening and closing of single ion channels (Figure 3.4). The information obtained in this way confirmed that ion channels were quite real and showed how they operated in response to the voltage changes that triggered action potentials and to the binding of neurotransmitters at synapses. When coupled with molecular genetic techniques, the method provided a way of eventually understanding the structure and function of many of these channels in a variety of cells. Within a few years, dozens of labs around the world were using this approach, and ion channels are now routinely studied in this way. When I was moderator of a retrospective many years later, I asked Katz whether there was anything in his remarkably long string of accomplishments that he wished he done differently. He answered without hesitation that he very much regretted not having invented the patch-clamp electrode, a relatively simple method that he could have pioneered, had he thought of it. But he was obviously pleased that one of his protégés had been partly responsible for the discovery, and he and Sakmann remained close until Katz's death in 2003.

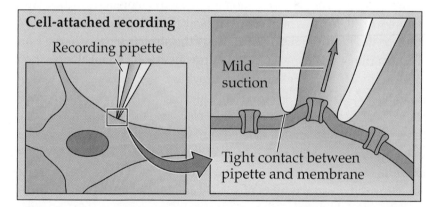

Cell-attached recording

Recording pipette

Mild suction

Tight contact between pipette and membrane

Figure 3.4   The technique developed by Bert Sakmann and Erwin Neher for measuring the activity of single ion channels (shown in the blowup). The method entails the application of a polished electrode to the membrane surface of nerve or muscle cells, enabling the recording of electrical events orders of magnitude smaller than the potential changes recorded with conventional electrodes that penetrate the cell membrane (see Figure 1.3). The diagram shows a small patch of membrane being gently sucked against the pipette, which is why the method is called "patch clamping." (After Purves, Augustine, et al., 2008)

After Katz stepped down as department head in 1978, Miledi was appointed his successor. Political problems apparently ensued both within the department and with the administration, and Miledi moved to the University of California at Irvine in 1985. The Department of Biophysics was disbanded a few years later, after 70 years of extraordinary productivity and a major impact on neuroscience.

During my last year at University College I also had to worry about getting a permanent job and what research I would pursue when I did. While still at Harvard in 1970, I had met Carlton Hunt when, as with Katz, he had come by to visit Kuffler (Hunt's middle name was Cuyler, and everyone called him Cuy) (Figure 3.5). Hunt, who was then in his early fifties, had been Kuffler's first fellow after Kuffler had arrived at Johns Hopkins from Australia in 1947, and had spent four years collaborating with him. Together they had worked on stretch receptors in muscle fibers (see Figure 3.5), a seemingly odd project but typical of Kuffler's nose for important and solvable problems in neuroscience. Most people take it for granted that the five senses (vision, audition, touch/pressure/pain, taste, and smell) provide all the fundamental information about the human environment that we need to survive in the world. In reality we possess dozens of other types of sensors that monitor what is happening within and around us, some of them more critical than the five obvious ones. One of the most important of these is the stretch sensors in muscles that continually inform the central nervous system about the position and status of the body's muscles, providing the feedback needed to maintain appropriate postures and perform successful motor acts. People who are blind or deaf get along reasonably well, but the absence of information arising from sensory receptors in muscles would be incompatible with life. No one had done much to explore how this sensory system works (the visual and auditory systems were, of course, the usual targets of such studies), and Hunt and Kuffler's collaboration had done a lot to put this key aspect of sensation in the prominent place it deserved.

When I first ran into Hunt at Harvard, he had recently moved to Washington University from Yale. Although I was already planning to go to England for another two years of training, I took note that Hunt was then in the process of building a Department of Physiology and Biophysics in St. Louis, having already put together excellent departments

A

B

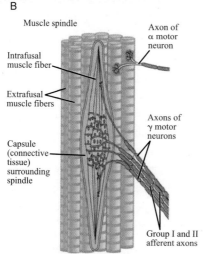

Figure 3.5 Cuy Hunt circa 1980, studying the properties of stretch receptors in muscles in his lab at Washington University. The diagram shows one of these complex sensors found in nearly all human muscles (the "extrafusal" fibers are the regular muscle fibers that generate body movements; the "intrafusal" fibers control the stretch receptor length to ensure ongoing sensitivity when muscles contract). Like Katz, Hunt fell into the bulldog category of scientists, and he continued to work on muscle spindles for more than 50 years. (Courtesy of Marion Hunt; the spindle diagram is from Purves, Augustine, et al., 2008)

at the University of Utah and then at Yale (where, as Chairman of Physiology, he had hired John Nicholls). Hunt was then—as always— a distinguished figure, and it was obvious that Kuffler and the rest of the faculty at Harvard liked him and admired the two departments that he had already created. Whatever conversation we had then about future plans must have been quite tentative. However, I took note that if history and first impressions were any guide, Hunt would be an excellent person to work for.

I met Hunt again two years later in summer 1972 when he visited Katz at University College. Hunt had done a sabbatical year in Katz's lab a decade earlier when he had taken a break from muscle spindles to study the effects on neurons of cutting their axons, which interrupts the connection between the nerve cells and their targets. He was a great admirer of Katz and took pains to visit whenever he was in England. Hunt took me to lunch at an upscale restaurant where we discussed what I had been doing and the possibility of coming to his new Department of Physiology and Biophysics. We agreed over coffee that I would visit St. Louis later that fall to have a look.

Although the trip in late October 1972 included the few other places that had indicated some interest in hiring me, I liked St. Louis, Washington University, and the potential colleagues I met. Washington University also had a rich history of research in neuroscience that was appealing. But most of all, I felt I would be comfortable working in a department run by Hunt, and that he could and would provide sound guidance to someone still relatively untutored in science and the ways of academia. The University of California at San Francisco was my other option, but in addition to all the other factors involved, Hunt's association with Kuffler and Katz promised continuity with the path I had started out on five years earlier.

And so, with some difficulty, I convinced Shannon (who had remained in London with our now 3-year-old daughter) that St. Louis was the right place for us—or for me, she would no doubt wish to add—and we arrived in the Midwest on a sweltering day toward the end of the following summer.

4

Nerve cells versus brain systems

It must already be apparent that, to an extraordinary degree, what one does in science is determined more by circumstances and chance than by guiding principles. This was certainly the case for me as I started life as a fully independent neuroscientist in St. Louis. Much of what I pursued depended on the colleagues in my new department and on the people who happened to cross my path. And this, in turn, depended on the organization of the faculty engaged in neuroscience at Washington University in 1973, which was fairly typical of American universities at the time.

Cuy Hunt's Department of Physiology and Biophysics was in the medical school adjacent to Barnes Hospital in the city proper; the rest of the campus was about 2 miles away, across the major city park in the more suburban setting of University City. The separation of medical schools from the rest of the university was common; in Boston, Harvard College in Cambridge had been even further removed physically and intellectually from Harvard Medical School. The reason primarily stemmed from the history of medical education. Medical schools in the United States in the eighteenth and nineteenth centuries had started as trade schools rather than the academic centers they have become. Therefore, they were located adjacent to hospitals, which were often far from the campuses of the universities that eventually came to run them. Washington University School of Medicine arose in this way, having been preceded by two private for-profit schools (St. Louis Medical College and Missouri Medical College) founded in the 1840s to train local physicians. Under the auspices of the university, the medical school incorporated these two proprietary schools in 1891, 20 years before Abraham Flexner's report that established curricular standards for U.S. medical schools nationally. The concept of medical schools as integral parts of research universities was even slower in coming and is still being resolved today with the construction of campuses that facilitate greater unity with undergraduate education. (As an undergraduate at Yale, I never set foot in the medical school, which offered no premed or other undergraduate courses.) The consequence of this geographical separation in St. Louis was that interactions between the faculty in undergraduate departments such as biology, psychology, and philosophy were less frequent than they should have been.

Even within the medical school, varying perspectives and tensions based on the history and traditions of different disciplines were apparent as I settled into my role as a junior faculty member. I had paid little attention to such issues as a postdoc at Harvard or University College London, but these matters were now relevant. In 1973, the integration of neuroscientists from such traditional departments as physiology, anatomy, or biochemistry into a single department of neurobiology was still unique to Kuffler's department at Harvard, which had been established in 1966. This evolutionary change, based on the growing importance of neuroscience, had not yet come to Washington University, so a significant factor in whom you were likely to discuss science with over lunch or in the hall depended very much on the department you were in. Medical students had to be taught the full range of physiology, so the Department of Physiology and Biophysics that Hunt put together included people who worked on the lungs, kidneys, and heart. Nonetheless, Hunt's particular enthusiasms clearly favored neuroscience, and 6 or 7 of the approximately 15 faculty members were neuroscientists. In addition to me (the newest member), the faculty included Carl Rovainen, who had been a graduate student with Ed Kravitz at Harvard and worked on the nervous system of the lamprey, a simple vertebrate with many of the advantages of invertebrates; Mordy Blaustein, who had been a postdoctoral fellow with Hodgkin at Cambridge and worked on the role of calcium ions in cell signaling; and Alan Pearlman and Nigel Daw, both of whom had been fellows in Hubel and Wiesel's lab at Harvard and were continuing to work on the visual system. Two others were not quite card-carrying neuroscientists but were close enough, according to the criteria of the day: Roy Costantin had trained with Huxley and worked on muscle contraction, and Paul DeWeer who worked on membrane pumps, the metabolically driven molecules in the membranes of neurons and other cells that generate the ion concentrations on which neuronal signaling depends. This group represented Hunt's inclination toward the subjects and people he had become familiar with as a fellow with Kuffler, his admiration for what Hodgkin and Huxley had accomplished, and the friendship with Katz that had developed during his sabbatical at University College. The faculty reflected the remarkable degree to which personal associations influence what goes on in academe.

A few floors away, Max Cowan's Department of Anatomy was the other important department doing neuroscience. Hunt had come from a background in physiology, but Cowan was a neuroanatomist in the tradition of Golgi, Cajál, and the other preeminent anatomists who followed in the first half of the twentieth century. Cowan had grown up in South Africa and had gone to the University of Witwatersrand, where he began studying medicine. In 1953, he transferred to Oxford at the invitation of British anatomist Wilfred LeGros Clark, where he completed his medical training and obtained a doctoral degree under Le Gros Clark's direction. From the outset, Cowan was interested in methods that could indicate the axonal connections between various brain regions and had carried out experiments with a series of collaborators in England that gained him a well-deserved reputation at an early age. In 1966, he took a faculty job at the University of Wisconsin. After only two years there, Washington University recruited him at the age of 37 to take over the moribund Department of Anatomy. By the time I arrived five years later, Cowan had already proved to be a brilliant choice. He had not only reinvigorated all the usual functions of the department, but had hired a cadre of outstanding young neuroanatomists (along with other good people needed to teach the gross and microscopic anatomy of the rest of the body). The neuroanatomists Cowan had recruited included Tom Woolsey, who worked on the somatic sensory system; Ted Jones, who was studying the connectivity of the thalamus; Harold Burton, who studied the organization of the somatic sensory system; Tom Thach, who studied the cerebellum; and Joel Price, who worked on the anatomy of the olfactory system. He had also recruited a couple cell and molecular biologists who shared his interest in axon biology and methods of tracing axonal connections. A handful of other basic neuroscientists were located in biochemistry and pharmacology, and in the clinical departments of neurology, neurosurgery, psychiatry, or radiology, but these people were less in evidence, and the latter tended to pursue disease-related research that was deemed of relatively little interest to basic research at the time. (In the five years that I had spent as a fellow at Harvard and University College, I don't recall a single seminar on a neurological disease; in sharp distinction to the situation today, the eventual relevance of basic research findings to clinical medicine was simply assumed.)

The two departments Hunt and Cowan ran reflected the distinct traditions of physiology (defined as the study of how cells and organ systems function) and anatomy (defined as the study of their structure). At the same time, both Hunt and Cowan were being driven by the logic of integration in neuroscience that Kuffler's department had been the first to formally realize. Both men saw themselves primarily as neuroscientists and had tilted their faculties strongly in that direction, although with a physiological bent in Hunt's case and an anatomical one in Cowan's. In recognition of the coming change, Cowan had already renamed his burgeoning enterprise the Department of Anatomy and Neurobiology, and considerable overlap between the departments and some competition between the two chairs were evident. Cowan asked me whether I would be interested in joining his department after I had been in St. Louis only a year or so. Although I declined the offer, that sort of poaching did little to improve the sometimes cool relationship between the two men.

The backgrounds, intellectual styles, and overall directions of the faculty in the two departments were characteristic of a dichotomy in neuroscience that persists today and, in some ways, has gotten worse. The difference was not simply whether one was more attracted to physiology or anatomy, but whether one was drawn to study the nervous system at the level of nerve cells and their synaptic connections or at the level of brain systems. The distinction between these different commitments had been apparent even in the nominally integrated Department of Neurobiology at Harvard, where Hubel and Wiesel, working on the visual system, had separated themselves physically—and, to a degree, intellectually—from the majority, who were working at a more reductionist level on various simple invertebrates or model systems. This divide is even more apparent now, due largely to the advent of molecular biology in neuroscience that began in the late 1960s and accelerated in the following decades. The enormous power of the new understanding of genes and the tools that were soon provided gave a big boost to both reductionists and clinician–scientists interested in neurological diseases, further emphasizing the differences between the neuronal and systems-level camps. Because studying how genes influence nerve cells, their interactions, and their role in various diseases is different than understanding how brain systems work, the gulf between reductionists and those seeking answers to questions about brain systems has widened. The problem has been further exacerbated by the shortsighted view of politicians, funding agencies, and university administrators who believe that research in neuroscience (and biology in general) should focus on human health.

Although my initial interests—philosophy, Freud, and psychiatry, when I graduated from college and started medical school—had been anything but reductionist, everything I had done for the preceding five years in neuroscience had been at a simple model level. The issues that Bert Sakmann and I had just been studying did not even involve nerve cells directly. I was ill prepared to launch into a project that focused on the structure and function of the brain, although I worried about whether the reductionist approaches I used could ever say much about the brain functions that had always seemed more interesting than cell and molecular interactions. But I was at least determined to work on nerve cells in a mammal as a step in the right general direction, and on problems that would have more pertinence to brain function and organization than the projects I had cut my teeth on.

How to do this was not obvious, but I had considered possibilities while I was in London. With Jack McMahan, I had worked on small collections of nerve cells in the peripheral nervous system called autonomic ganglia that several of Kuffler's collaborators were working on at the time. Studying the function and organization of these accessible collections of neurons that have connections with both the central nervous system and peripheral targets seemed like a good compromise between plodding onward with some aspect of a model synapse, such as the neuromuscular junction, and taking a more direct attack on some aspect of the brain, which I knew very little about at that point. Figure 4.1 illustrates how the autonomic nervous system in humans and other mammals controls a wide range of involuntary functions mediated by the activity of smooth muscle fibers, cardiac muscle, and glands. The system comprises two major divisions. The sympathetic component of the system mobilizes the body's resources for handling biological challenges. In contrast, the parasympathetic system is active during states of relative quiescence, enabling the restoration of the energy reserves previously expended in meeting some demanding contingency. This ongoing neural regulation of resource expenditure and replenishment to maintain an overall balance of body functions is called homeostasis. Although the major controlling centers for homeostasis are the hypothalamus and the circuitry it controls in the brainstem and the spinal cord, the neurons that directly activate the smooth muscles and glands in various organs are in collections of hundreds or thousands of nerve cells in the autonomic ganglia shown in Figure 4.1.

Autonomic ganglia had been the focus of many key studies of the nervous system since the middle of the nineteenth century. But despite its technical advantages and physiological importance, the autonomic system had always been regarded as a relatively inferior object for research compared to components of the mammalian brain that, understandably, attracted more interest—the visual system, the auditory system, the somatic sensory system, and the voluntary (skeletal) motor system, in particular. Although humans must long ago have observed involuntary motor reactions to stimuli in the environment (such as the pupils narrowing in response to light, superficial blood vessels constricting in response to cold or fear, and heart rate increasing in response to exertion), the neural control of these and other visceral functions was not understood in modern terms until the late nineteenth century. The researchers who first rationalized the workings of the autonomic system were Walter Gaskell and John Langley, two British physiologists at Cambridge. Gaskell, whose work preceded that of Langley, established the overall anatomy of the system and carried out early physiological experiments that demonstrated some of its salient functional characteristics (showing, for example, that the heartbeat of an experimental animal is accelerated by stimulating the outflow to the relevant sympathetic ganglia and slowed by stimulating the outflow to the relevant parasympathetic ganglia; see Figure 4.1). Based on these observations, Gaskell concluded in 1866 that "every tissue is innervated by two sets of nerve fibers of opposite characters" and further surmised that these actions showed "the characteristic signs of opposite chemical processes."

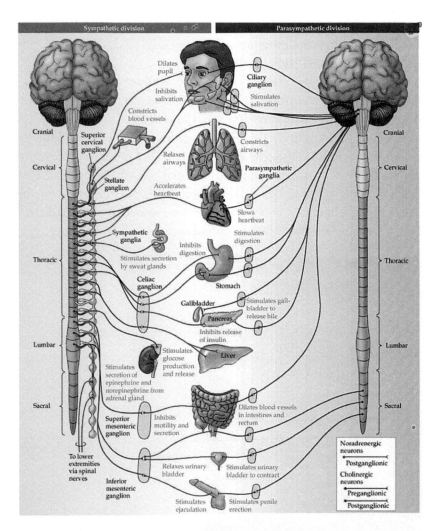

Figure 4.1 The autonomic nervous system, which controls the body's organ systems and glands. These homeostatic functions, much like the information that arises from the muscle sensors described in the last chapter, are critical to survival. Many of the neurons involved are in clusters called autonomic ganglia that lie outside the brain and spinal cord (see Figure 1.6), making it relatively easy to study their organization and function. (From Purves, Augustine, et al., 2008)

Langley (Figure 4.2) was the real giant in this aspect of neuroscience. He established the function of autonomic ganglia, coined the phrase *autonomic nervous system,* and carried out studies on the pharmacology of the autonomic system that eventually established the roles of acetylcholine and the catecholamines, the first of the many neurotransmitter agents that were identified in the early decades of the twentieth century. This work set the stage for understanding neurotransmitter action at synapses and, ultimately, for Katz's discoveries of the detailed mechanism of chemical synaptic transmission. Langley's work also led to studies of the autonomic nervous system by Walter Cannon at Harvard Medical School. Among many other accomplishments, Cannon established the effects of denervation during the 1940s, laying the foundation for Miledi's work in the 1960s on the spread of sensitivity when muscle fibers are deprived of their innervation; that, in turn, set up the project that Sakmann and I worked on at University College. For better or for worse, this skein of personal and intellectual associations is characteristic of the way research themes unfold in any branch of science.

Figure 4.2   John Langley, a central figure in late-nineteenth-century neurophysiology and a pioneer in studies of the autonomic nervous system and synaptic transmission by chemical agents.

Whatever the merits of choosing to work on the autonomic system, getting started in St. Louis depended on much more than simply picking a reasonable topic to study. Hunt was a great help, and I soon understood why he had attracted such good people to the three different departments that he had organized, and why their research generally flourished. He was every bit the paternal adviser that I had imagined, and he helped me get going in all kinds of ways, many of them having nothing to do with science. After I had been plugging away for a couple of months with the equipment that he had arranged to have waiting for me in St.Louis (new, and much finer than what I had been used to working with during the preceding two years at University College), he came by the lab one day to ask why I had chosen not to enroll in TIAA-CREF, the academic pension fund. I told him that I really wasn't worried about retirement at that point, and that Shannon and I couldn't afford to pay the monthly contribution. He patiently explained what an annuity was, extolled the virtues of compound interest, and told me why this would eventually be important. More to the point, he raised my salary that very day so that we could afford to make the contribution.

After I had been working in St. Louis for about six months, Hunt again wandered into the lab one afternoon and asked me in an offhand way whether I had ever thought of applying for a research grant. When I said, with extraordinary naivete, that I really didn't understand how research grants worked, he explained that science costs money and that the people who do it are expected to raise the funds to pay for it. He did all this in a way that made my ignorance seem perfectly okay. With his editorial help, I soon had a research grant. For the following 11 years that I worked in Hunt's department, he never suggested to me—or, as far as I know, to anyone else—what to do or how to do it, although he was always happy to talk about the science and offer his expertise, which was considerable. Promotions happened as if by magic, although I now know that Hunt, who died in 2008 at the age of 89, had to put together dossiers, solicit letters of recommendation, and convince a cantankerous committee of fellow department chairs that a faculty member was worth advancing. He did all this while he successfully pursued his own work on the sensory physiology of muscle spindles, demonstrating by example that being an administrator does not mean relinquishing good science. It no doubt helped that, during all those years, Hunt—to the best of my memory—convened only a single faculty meeting.

Working on neuronal connections in the mammalian autonomic system turned out to be a good choice. Although I saw this work as a stepping stone toward a more direct attack on problems explicitly related to brain function, the step eventually consumed about a dozen years with results that, in contrast to what I had done up to that point, were regarded as important. During the first few years in St. Louis, I undertook two projects in the peripheral autonomic system of mammals. The first was directly inspired by observations Langley had made 80 years earlier. In the course of brain development in embryonic and early postnatal life, connections between nerve cells must be made appropriately and not just willy-nilly, a process referred to as neural specificity. Experience later in life is important in the ultimate organization and further refinement of connections (see Chapter 3), but the idea that the human brain or any other brain comes into the world as a *tabula rasa* to be imprinted primarily by knowledge derived from experience—the model suggested by eighteenth-century philosopher John Locke—is silly. Nervous systems at birth are already connected in detailed and highly specific ways based on the experience of the species over evolutionary time. The mechanisms that produce this specificity of connections during development were unclear in 1973 and still aren't fully known today.

Langley examined this issue at the end of the nineteenth century, making use of the fact that neurons at different levels of the spinal cord innervate neurons in sympathetic ganglia in a stereotyped way (Figure 4.3). In the superior cervical ganglion, for example, cells from the highest thoracic level of the spinal cord (T1) innervate ganglion cells that project to smooth muscles that dilate the pupil, whereas neurons from a somewhat lower level of the cord (T4) innervate ganglion cells that cause effects in other targets, such as muscles that constrict blood vessels in the ear. Langley had assessed these differences in the innervation of the ganglionic neurons simply by looking at these peripheral effects while electrically stimulating the outflow to the ganglion from different spinal levels in anesthetized cats, dogs, and rabbits. When he stimulated the outflow from the upper segments of the thoracic spinal cord, the animals' pupil dilated on the stimulated side without any effect on the blood vessels of the ear. And when he stimulated the lower thoracic cord segments, the pupils were not affected, but the blood vessels in the ear on that side constricted. When he cut the sympathetic trunk that carried the axons to the ganglion and waited some weeks for them to grow back, he observed the same pattern of peripheral responses. Therefore, Langley surmised that the mechanisms underlying the differential innervation of the ganglion cells must occur at the level of synapse formation on the neurons in the ganglion, and further suggested that selective synapse formation is based on differential affinities of the pre- and postsynaptic elements arising from some sort of biochemical markers on their surfaces.

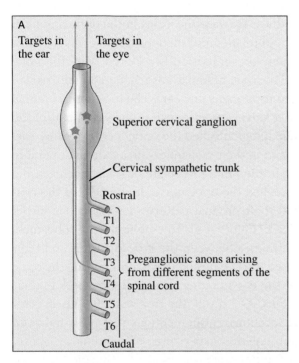

A

Targets in the ear

Targets in the eye

Superior cervical ganglion

Cervical sympathetic trunk

Rostral

T1

T2

T3

T4

T5

T6

Preganglionic anons arising from different segments of the spinal cord

Caudal

Figure 4.3 Specificity of synaptic connections in the autonomic nervous system. This diagram shows the superior cervical ganglion and its innervation by neurons located in the thoracic portion of the spinal cord; the ganglion is in the neck (see Figure 4.1). Langley used this part of the mammalian sympathetic system to demonstrate indirectly that neurons in the ganglion can distinguish between axon terminals arising from different levels of the spinal cord as synaptic connections are being made. (After Purves, Augustine, et al., 2008)

Between the 1940s and the early 1960s, Roger Sperry carried out experiments that led to a modern articulation of what is now called the chemoaffinity hypothesis. Sperry was an equally remarkable neuroscientist whose long career unfolded mainly at Cal Tech, where he later worked on the functional specialization of the right and left cerebral hemispheres that won him even greater acclaim. (Most fundamental work on left–right brain differences—and many nonsensical New Age ideas on this subject—can be traced to Sperry's discoveries on patients whose hemispheres had been surgically separated to treat epilepsy.) His conclusions on neural specificity were based on studies similar in principle to Langley's, but carried out in the brains of frogs and goldfish instead of in the peripheral nervous system. In humans and other mammals, damage to the optic nerve causes permanent blindness because the axons in the optic nerve fail to regenerate. But in amphibians and fish, optic axon nerves regenerate after they have been cut, and vision is restored (why the optic nerve and other central nervous system axons regenerate quite well in these animals but not in us remains unclear). The terminals of retinal axons normally form a relatively precise map in the visual part of the fish or amphibian brain, a region called the optic tectum. Axons arising from a particular point in the retina innervate a particular point in the tectum, preserving in the brain the neighbor relationships in the retina. When Sperry crushed the optic nerve, he found that the retinal axons reestablished their original pattern of connections in the tectum as they grew back (Figure 4.4). To emphasize the robustness of the specific chemoaffinities between the growing axons and their target neurons, he turned the eye upside down after cutting the optic nerve and showed that the regenerating axons still grew back to their original tectal destinations. As a result, the frog was left with an erroneous sense of object locations, misperceptions that persisted even after months of subsequent experience. Accordingly, Sperry proposed that each cell in the brain carries an identification tag, and that growing axons have complementary tags that enable the axons to seek out and contact specific neurons.

Given these studies by Langley and Sperry, it seemed worthwhile to pursue the issue of neural specificity at the level of electrical recordings from individual neurons in autonomic ganglia. Working with Arild Njå, a postdoctoral fellow from Oslo who was the first to come my way, we pursued the merits of this idea in the autonomic system of guinea pigs by removing the whole upper portion of the sympathetic chain (shown in Figure 4.3), keeping it alive in a chamber, and making intracellular recordings from individual neurons in the superior cervical ganglion while stimulating each of the input levels from the spinal cord. The results showed that the synaptic connections made on ganglion cells by preganglionic neurons of a particular spinal level are

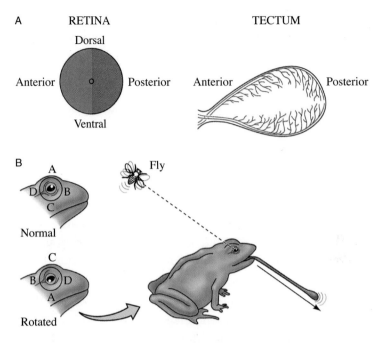

Figure 4.4 Roger Sperry conducted experiments on the specificity of synaptic connections in the brain more than 60 years after Langley's work, making the same point in the visual system of the frog. When the axons of neurons in the retina grow back to the part of the brain called the optic tectum, they primarily contact the same nerve cells that they did initially. As a result, when the axons regrow after the eye is rotated, the frog's brain provides wrong information about the location of objects in the world. (After Purves, Augustine, et al., 2008)

indeed preferred, but that contacts from neurons at other levels are not excluded. Furthermore, if the innervation of the superior cervical ganglion was surgically interrupted, recordings made some weeks later indicated that the new connections again established a pattern of segmental preferences. Therefore, spinal cord neurons associate with target neurons in the autonomic ganglia of mammals according to a continuously variable system of preferences during synapse formation that guide the pattern of innervation during development or reinnervation without limiting it in any absolute way.

Although this work with Njå· resulted in several good papers, it was another project I had begun in parallel that eventually occupied most of my attention during the next decade. The ideas on which this work was based came from a different direction and again demonstrate the importance of proximity and happenstance. The theme that

Sakmann and I had been working on in London was control of the signaling properites of neurons (although we used muscle cells as a model), and I continued to think—along with many other neuroscientists—that such modulation of signaling and its effects on long-term connectivity were especially important. Hunt and I had discussed his work on the changes of neuronal properties that occur when a neuron's axons are cut, the issue that he had worked on during his 1962 sabbatical in Katz's lab. And I knew that Cowan had used anatomically visible changes in neurons when their axons are severed as a means of tracing axonal pathways in the brain. I was also aware that two neuroantomists at Oxford, Margaret Matthews and Geoff Raisman, had recently published a paper describing changes in the appearance and number of synapses made on superior cervical ganglion cells after cutting the connections of these neurons to their peripheral targets.

This evidence that a neuron's connection to its target was affecting how other nerve cells made synaptic connections with it seemed like a good topic to pursue, and so I did. In what turned out to be the first couple of papers to come out of my lab in St. Louis, I showed by electrophysiological recording that the efficacy of the synapses made by the spinal neurons on the neurons in the superior cervical ganglion declined during the first few days after the axons from the neurons to peripheral targets in the head and neck had been cut, and that this decline occurred in parallel with the loss of a majority of the synapses made on the ganglion cells that could be counted in the electron microscope (Figure 4.5). Because the loss of synapses from the neurons was reversed when the axons grew back to their peripheral targets, the conclusion seemed clear: The synaptic endings made on nerve cells must be actively maintained. And whatever the mechanism, this maintenance depended on the normal connections between nerve cells and the targets that they innervated. The clarity of these results in a relatively simple system of mammalian neurons was news and encouraged me to study these issues further.

This research led to the beginning of a long collaboration with Jeff Lichtman and a deepening friendship with Viktor Hamburger, both of whom turned out to be critical in determining how this work would progress. Jeff (Figure 4.6) appeared in my lab one day in 1974 and asked whether he could chat about his future. He was then a

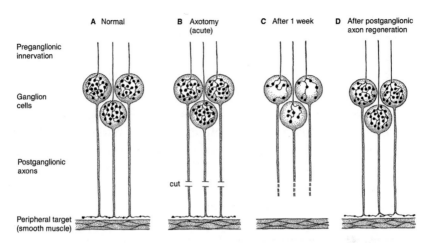

A Normal　　**B** Axotomy (acute)　　**C** After 1 week　　**D** After postganglionic axon regeneration

Preganglionic innervation

Ganglion cells

Postganglionic axons

cut

Peripheral target (smooth muscle)

Figure 4.5　Diagram illustrating the dependence of synapses on nerve cells in autonomic ganglia on the connections of these neurons to peripheral targets. When this link to the targets (such as smooth muscles in the eye and ear) is interrupted by cutting the peripheral (postgangionic) axons, most of the synapses on the nerve cells are lost; however, when the cut axons grow back to their targets, the synapses on the ganglionic neurons are restored. (Reprinted by permission of the publisher from *Body and Brain: A Trophic Theory of Neural Connections* by Dale Purves, p. 103, Cambridge, Mass: Harvard University Press, Copyright © 1988 by the President and Fellows of Harvard College.)

second-year med student and knew me from the lectures on neural signaling that I had given to his class some months before. Jeff was one of the ten or so med students in his cohort in the M.D./Ph.D. program, and he was trying to figure out what to do for his doctoral work. (The students in this ongoing federally subsidized program in the United States typically spend about four years doing basic research in addition to the four years of med school. This education is meant to generate physicians who can better bridge the gap between clinical medicine and basic science.) Jeff seemed nervous and lacked any good reason for wanting to work with me or ideas about what to do. I think he simply saw me as someone who was young and ambitious, and who, based on the lectures he had heard, might be a good mentor. My inclination was not to take him on because my experience at Harvard and University College had been that the best people populated their labs with postdoctoral fellows and not graduate students (neither Kuffler nor Katz had any graduate students when I worked in their departments; Hubel and Wiesel, whom I admired greatly, were likewise fellow oriented). But before reaching a decision, I thought it would be a good idea to ask Hunt for his thoughts on the matter. He pointed out that the M.D./Ph.D. students were a highly select group, that Lichtman would not cost me anything because the program was fully funded by the National Institutes of Health, and that unless I had a very good reason not to, I should certainly take him on as a graduate student. Hunt was indeed right: Lichtman was—and remains—one of the smartest and most imaginative people I have known in neuroscience. A decade after we had this conversation, Hunt hired him as a faculty member and Lichtman went on to become a major figure in neuroscience.

Figure 4.6   Jeff Lichtman circa 2006. (Courtesy of Jeff Lichtman)

Getting to know Hamburger was equally important, but this happened only gradually during the next few years. Despite the considerable scientific accomplishments of Hunt and Cowan, Hamburger was far and away the most notable neuroscientist at Washington University in 1973. Because he was in the Biology Department on the undergraduate campus, I had not met him on my trip to St. Louis as a faculty candidate; to my great embarrassment, I knew little or nothing about him or his work when I arrived in St. Louis. When I first bumped into Hamburger, I mistook him for another neuroscientist named Hamberger, a researcher of no great distinction who had studied the anatomy of the autonomic system. This woeful ignorance illustrated the parochial nature of my training and exposure up to that point. Hamburger was a consummate biologist, and my conversations with him about his work and the broader field of neural development over the next few years made me think much more about what nervous systems do for animals and less about the details of neurons.

5

Neural development

I first became aware of Viktor Hamburger at the seminar series that Max Cowan held every Saturday morning. Most of the hard-core neuroscientists at Washington University from both the medical school and the undergraduate campus made a point of attending these sessions, and Hamburger (Figure 5.1) was a regular. Initially, he was simply the old guy from the Biology Department who usually sat toward the back of the room. (He was then 73.) When he asked a question or made a comment in precise but heavily accented English, however, it was always authoritative and smart. Eventually, I asked around about him. After I figured out that he was a major figure in the field of neural development and not the neuroanatomist I had confused him with, and he figured out that I was a new hire in Hunt's department working on a problem he knew something about, we struck up an acquaintance. It grew into an important scientific and personal friendship that lasted until his death in 2001, more than 27 years later.

Hamburger had been a faculty member in the Biology Department across the park from the medical school since 1935 and was an enormously knowledgeable and accomplished neuroscientist, in the same league with Hodgkin, Huxley, Katz, and Kuffler. But he was working in a different area that did not overlap with research on neural signaling. Although it soon became clear that his work was pertinent to the research that I was starting, Hamburger's fundamental studies of how the nervous system develops had not been part of the traditional neurophysiology and neuroanatomy that I had been exposed to at Harvard or University College.

Figure 5.1   Viktor Hamburger in his office in the Biology Department at Washington University in the late 1970s.

Hamburger was born in 1900 and grew up in a rural town in Silesia, then a province of Germany (and now part of Poland). He attended the universities in Breslau, Heidelberg, and Munich before getting his Ph.D. under Hans Spemann at the University of Freiburg in 1925 (moving around academically was not usual in that era). Spemann, whose work was equally unknown to me, came from a pedigree of German embryologists and zoologists as distinguished as the British physiologists and émigrés such as Katz and Kuffler who had joined them in the 1930s. Spemann had won a Nobel Prize in 1935 for his studies of embryonic development and had trained a generation of leading embryologists, Hamburger preeminent among them. After a few years as a junior faculty member in Germany, Hamburger had come to the United States in 1932 under the auspices of a Rockefeller fellowship (the same fellowship that had sponsored Katz's emigration to England a few years later) for what was to have been a year working with Frank Lillie, an American embryologist who was studying the development of chicks at the University of Chicago. During that year, Hamburger received a letter dismissing him from his position at the University of Freiburg because he was Jewish. As a result, he stayed in Chicago for three years until he was hired at Washington University. In St. Louis, he had continued turning out a series of key studies on the development of the nervous system of chicks, a preparation he continued to explore until he quit doing experimental work more than 50 years later.

As I gained some familiarity with Hamburger's work, it dawned on me that his expertise was especially relevant to what I was doing in the autonomic system, or at least to the part that involved looking at the failure of synaptic maintenance when neurons were cut off from their peripheral targets (see Chapter 4). Hamburger had used the embryonic transplantation techniques he learned from Spemann to either add or remove limb buds (the forerunners of the chicken's wings and legs) in embryonic chicks, assessing what happened to the spinal neurons that would have innervated the ablated limbs, or that innervated the extra peripheral targets (Figure 5.2). By making a

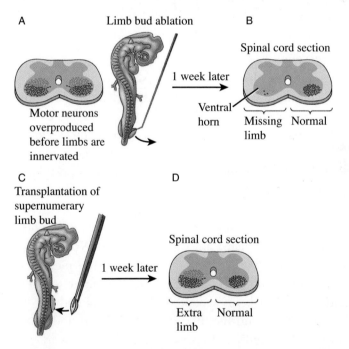

Figure 5.2   Hamburger's experiments showing how the presence or absence of peripheral targets affects the number of related motor neurons in the developing spinal cord of chicks. A, B) Amputation of the limb bud in an early embryo depletes motor neurons in the relevant region of the cord examined microscopically after the chick has hatched. C, D) Conversely, implanting an extra limb bud augments the amount of muscle and other tissue in the periphery, increasing the number of related spinal motor neurons. (After Purves, Augustine, et al., 2008)

window in the egg, he could carry out these operations in embryos at nearly any stage of development with the skillful use of fine glass needles. Hamburger found that when he examined the spinal cord microscopically after the chicks had hatched, the number of corresponding motor neurons in the spinal cord was much diminished if the progenitor of the hind limb had been amputated at an early embryonic stage (Figure 5.2A). Conversely, if he implanted an extra limb bud, the hatched chicks had more motor neurons than normal in the related part of the spinal cord (Figure 5.2B). Evidently the amount of peripheral target tissue the axons of spinal neurons encountered when they grew out into the limb bud somehow regulated the number of developing motor neurons in the spinal cord. This work begun in the 1930s established the phenomenon of target-dependent neuronal death or survival. By the 1970s, neuronal death regulated in this way during development was known to be a general phenomenon in the peripheral nervous system and in some parts of the central nervous system as well.

Hamburger's early results implied the presence of some agent in the targets of the axons growing into the limb bud that instructed the axons and their parent neurons about how much tissue existed peripherally and the degree to which they had been successful in making contact with it. A corollary was that the developing neurons in the spinal cord were competing for acquisition of the postulated material, dying off if they didn't get enough of this trophic agent (*trophic* refers to nourishment, and this word described what was evidently happening). As in most science, recognition of these implications occurred more slowly than I am suggesting here: Not until the 1950s, when much more work had been done, was the picture as clear as I am making it seem. But by the time I met Hamburger, these discoveries had long since been accepted. Although nobody in my Department of Physiology and Biophysics was very interested in any of this, Cowan was thoroughly familiar with Hamburger's work and had based many of his own studies of the developing chick brain on it.

Understanding how trophic interactions regulate neuronal numbers in early development had been greatly advanced by a collaboration that Hamburger began with Rita Levi-Montalcini in the late 1940s. Their work together lasted for about eight years and led to the discovery of nerve growth factor, a trophic molecule derived from smooth muscle that is the peripheral "nourishing" agent for at least two types of neurons, one of which was the nerve cell type in the ganglia of the sympathetic nervous system I had been working on. Nerve growth factor has served as a paradigm for the interaction between nerve cells and their targets ever since, and it remains the central example of trophic interactions in neurobiology (ironically, the factor that mediates the trophic effects of skeletal muscle on the spinal neurons shown in Figure 5.2 has, despite much effort, never been identified). Among other reasons for the intense ongoing interest in trophic factors is the possibility that neurological diseases that entail neuronal death (such as Alzheimer's disease and amyotrophic lateral sclerosis) are disorders of trophic interactions, and that better understanding this aspect of neurobiology could provide useful treatments. Many millions of research and biotech dollars have been invested in this idea during the last 30 or 40 years, although so far without clinical success. And although other growth factors are clearly important in brain development and function, that story has turned out much more complex.

The collaboration began in 1946 when Hamburger wrote to Levi-Montalcini in Turin to ask her if she would be interested in working in his lab for a year. The invitation made good sense because Levi-Montalcini had also been studying cell death and the support of developing neurons by peripheral targets. She accepted Hamburger's invitation and remained in the Department of Biology at Washington University for more than 25 years, a period that she described as the "happiest and most productive of my life." She returned to Rome in 1974 to continue her research there, but was a frequent visitor to Washington University, where she retained a part-time appointment. By any criterion, Levi-Montalcini was (and still is, at the age of 100) a remarkable scientist and a heroic figure. When she graduated from medical school in Turin in 1936, she had intended to pursue a clinical career in neurology and psychiatry, but that plan had to be put aside in 1938 by Mussolini's prohibition of "non-Aryans" from academic positions. Although Levi-Montalcini could have emigrated, she chose to remain in Italy, carrying out research in a small laboratory she set up in her home in collaboration with neuroanatomist Guiseppe Levi, her professor and mentor when she was a medical student (they were not related). Despite these conditions, she and Levi, who had been dismissed from his post because he was Jewish, managed to turn out several important papers during the war, stimulated in part by Hamburger's earlier studies on cell death. Hamburger's admiration of this work led to the invitation he tendered.

Hamburger and Levi-Montalcini's discovery of nerve growth factor and the subsequent isolation of the molecule by Levi-Montalcini and biochemist Stanley Cohen introduced a whole new perspective about how nerve cells interact with each other and with non-neural targets. The long path to this discovery began with one of Hamburger's students in the 1940s, a little-known figure in the history of neuroscience named Elmer Bueker. As illustrated in Figure 5.2B, one of the issues that had interested Hamburger was augmentation of the periphery as a means of salvaging developing neurons that would otherwise die during the normal course of development. Shortly after getting his Ph.D. with Hamburger, Bueker (who had taken a position at Georgetown University) had the unusual idea of implanting tumor tissue into a chick limb bud as a means of augmenting the periphery in a simpler and more dramatic way. Although the tumor that worked best had little or no effect on the neurons in the spinal cord, he noted that the sensory and autonomic ganglia (see Figure 4.1) were obviously enlarged on the side of the chick embryos in which the tumor was implanted. Bueker showed this result to Hamburger and Levi-Montalcini in 1948. They quickly appreciated its potential significance, and although Bueker went on to an unremarkable career as an anatomist at Georgetown and then New York University, Hamburger and Levi-Montalcini began to pursue his observation with a vengeance.

Their enthusiasm was fueled by the possibility of extracting and ultimately identifying the active agent from Bueker's tumor tissue. While spending a few months at the Carlos Chagas Institute in Brazil in a lab that specialized in tissue culture, Levi-Montalcini devised a way of assaying the presumptive agent in the tumor by measuring the outgrowth of nerve cell processes (axons and dendrites) from embryonic chick ganglia placed in tissue culture medium laced with tumor extract (Figure 5.3). This exuberant outgrowth of neuronal processes had caused the gross enlargement of the ganglia Bueker had first observed. Stanley Cohen, then a postdoctoral fellow in the Department of Biochemistry at the school of medicine, was invited to join the Hamburger lab to help solve the problem. Although they initially worked together in the early 1950s, Cohen and Levi-Montalcini continued without Hamburger, using this assay during the mid-1950s to isolate and eventually identify what they called nerve growth stimulating factor. (The molecule was eventually sequenced by another group at Washington University in 1971 and for decades has simply been called nerve growth factor, or NGF.)

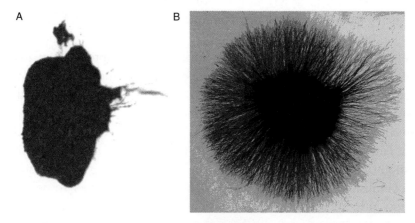

Figure 5.3   The bioassay that enabled identification of nerve growth factor. A) A chick ganglion grown in tissue culture for several days without nerve growth factor. B) Another ganglion grown for several days with the factor, showing the extraordinary outgrowth of nerve cell processes stimulated by this agent. (After Purves and Lichtman, 1985)

Despite its importance in understanding how neural circuitry is normally established and maintained (Levi-Montalcini and Cohen went on to win the Nobel Prize in Physiology or Medicine for this work in 1986), the larger community of neuroscientists did not immediately appreciate this work. I first encountered Levi-Montalcini in 1970 when she visited Kuffler's department at Harvard to give a lunchtime seminar. She talked about studies she had been carrying out on the effects of nerve growth factor on cockroach neurons, a tangent she undertook for reasons I don't remember. When she continued speaking well beyond the allotted hour and was only halfway through a second carousel of slides, Kuffler surreptitiously motioned his lab technician, who was running the projector, to advance the carousel. The talk finished promptly, apparently without

Levi-Montalcini noticing the elision of 20 or so remaining slides. The talk didn't seem to make much of an impression on anyone, least of all me. She didn't mention the importance of Hamburger's work in her exposition, explaining in part why he was unknown to me when I got to St. Louis a few years later.

Although Hamburger described her as having been "mousey" when she arrived in St. Louis in 1947, by the time I came in 1973, Levi-Montalcini was a distinctly regal figure who contrasted sharply with Hamburger's description of her 25 years earlier (Figure 5.4). (Having been long divorced, Hamburger was a man who noticed such things, and he had devoted girlfriends until he outlived them all well into his nineties.) Although Hamburger had opted out of the effort to isolate and identify nerve growth factor in the 1950s when it became clear that the enterprise had become largely biochemical, the Nobel Committee unfairly excluded him from the award presented to Cohen and Levi-Montalcini in 1986. Levi-Montalcini was largely responsible for the exclusion. The ambition and determination that had enabled her to pursue research during the war and accomplish so much in the face of long odds also led her to politic incessantly for recognition as a preeminent figure in neuroscience, which she certainly was. In this quest, she diminished Hamburger's importance in their collaboration, even after she had triumphed in Stockholm. After the prize had been awarded to her and Cohen in 1986, my colleague Josh Sanes and I wrote an article praising their accomplishment, but pointing out that Hamburger had been treated shabbily because his earlier work, his invitation to Levi-Montalcini, the results of his student Bueker, and Hamburger's collaboration through the early 1950s had obviously been critical to their success. As a matter of routine, the journal in which the article was to appear had shown the proof to Levi-Montalcini to check the facts. The next day, I was awakened by an angry phone call from Levi-Montalcini in Rome complaining that the article gave Hamburger credit that he did not deserve and asked us to revise it. Sanes and I did not change the article, but the call made her conflicted feelings abundantly clear. Hamburger never understood this behavior toward him, and although they continued a polite and superficially cordial relationship, he never forgave her.

Figure 5.4 Rita Levi-Montalcini in 1977. (From Purves and Lichtman, 1985)

What I learned from Hamburger about neural development and nerve growth factor during the next few years had a significant impact on what was going on in my lab, where people were toiling away on the formation and maintenance of synaptic connections in the simple and accessible systems that various autonomic ganglia in mammals provided. By then, Levi-Montalcini was spending most of her time in Rome, despite her appointment in St. Louis, and my interactions with her were mostly on social occasions with Hamburger. Like Hamburger, I had never had much interest in studies at the molecular level; among other reasons, my brief but dismal experience with neuropharmacology research as a med student left a lingering bad taste, and, with occasional striking exceptions (such as the discovery of endorphins in the 1970s), I felt that many molecular studies were revealing more and more about less and less. Nerve growth factor was another exception. Not only did this agent promote the survival of the very neurons we were studying, but it also influenced the growth of the axonal and dendritic processes of the classes of neurons that were sensitive to it and, by implication, the synaptic contacts they made (see Figure 5.3). It was not much of a stretch to imagine that competition for and acquisition of such factors was the basis of the maintenance of synaptic connections we had been providing evidence for, and that this "trophic theory" of how synapses were regulated in the nervous system was a general rule. The idea was that each class of cells in a neural pathway was supporting and regulating the connections it received by trophic interactions with the cells it innervated down the line, resulting in a coordinated chain of connectivity that extended from the periphery centrally to the spinal cord and, ultimately, on through the controlling centers in the brain (Figure 5.5).

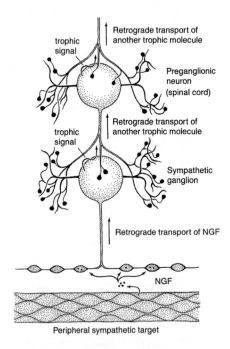

Figure 5.5   Scheme of synaptic maintenance that could enable the coordination of synaptic connectivity throughout an entire neural pathway (NGF stands for nerve growth factor). (Reprinted by permission of the publisher from *Body and Brain: A Trophic Theory of Neural Connections* by Dale Purves, p. 135, Cambridge, Mass: Harvard University Press, Copyright © 1988 by the President and Fellows of Harvard College. After Purves, 1986.)

The goal of this work on synaptic connectivity in mammals was not to sort out what molecules might be involved (the paradigm provided by nerve growth factor was sufficient evidence, and many labs were studying this agent by the mid-1970s), but instead to determine the governing principles. Jeff Lichtman was the prime mover in pursuing this aim. My initial doubts about taking him on as an M.D./Ph.D. student had been quickly dispelled. Within a few weeks, it was obvious that Lichtman was extremely bright, and even though he was a student 13 years my junior, we were soon discussing things as colleagues. Persistence is another good quality for a scientist to have, and Lichtman was as tenacious and determined as he was smart. Shortly after he started working in the lab, I found him underneath one of the cabinets removing the drain trap with a wrench. He had stained some neurons in a ganglion with a new dye technique, and the ganglion had accidentally gone down the sink during the procedure. Although the missing ganglion was only as big the head of a pin, Jeff eventually found it in the muck that he pulled out of the trap.

The main problems that concerned us for the next several years were the nature of competition among the axons that innervate target nerve cells, and how the signaling activity of competing nerve cells affects the balance of synaptic connectivity (a problem related to the effects of activity that Bert Sakmann and I had wrestled with in London). We were convinced that nerve cells and their targets must interact in sorting out the connectivity of functioning circuits in much the same way that elements in an ecosystem eventually establish equilibrium as they compete for limited resources. This idea about neural connectivity was not new—the Spanish neuroanatomist Ramon y Cajál had written in flowery prose about this ecological concept of neural development in the late nineteenth century—but no one had determined the neurobiology of how this competition might actually work.

The closest anyone had come to directly exploring the issue of synaptic competition by the mid-1970s was work by Michael Brown, David Van Essen, and Jan Jansen on the developing innervation of skeletal muscle fibers. I didn't know Brown, but Van Essen had been a graduate student at Harvard when I was there (he joined John Nicholls's lab for his doctoral work about the time I left to work with Jack McMahan), and Jansen had worked in Nicholls's lab when on sabbatical from his position in Oslo that same year. A few years later in Oslo, they had shown that during the first few weeks of postnatal life, each fiber in a rat muscle is contacted by more nerve terminals from different axons than persist in maturity (Figure 5.6A), providing

another clue about the nature of synaptic competition and maintenance. A natural question was whether the innervation of neurons
followed the same rules as muscle fibers, and Lichtman's thesis work
showed that it did (Figure 5.6B).

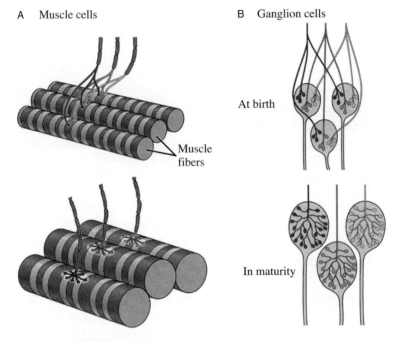

A   Muscle cells   B   Ganglion cells

Muscle
fibers

At birth

In maturity

Figure 5.6   The competitive interaction between axon terminals for the innervation of target cells during synaptic development. A) The elimination of all
synapses except those made by a single axon on each developing muscle fiber
during the course of early postnatal life. B) The analogous phenomenon on
maturing neurons (in this case, on a class of autonomic ganglion cells that
lack dendrites, making the analysis much simpler). (Adapted from Purves,
Augustine, et al., 2008)

Understanding the interactions among axon terminals and the
synapses they make on target cells remains woefully incomplete
today. However, some important principles emerged from work that
Lichtman and several other fellows carried out in the lab over the
next few years. One principle is that the spatial configuration of a
neuron is a critical determinant of the innervation it receives. For
nerve cells without dendritic processes, such as those in Figure 5.6B,
the end result of the initial competition is innervation by many synaptic endings that all arise from the same nerve cell axon. If target nerve

cells have dendrites, however, the number of innervating axons increases in proportion to the number and complexity of these branches (Figure 5.7A). Moreover, after a given axon makes some synapses on a target neuron, the axon is somehow informed by the conjoint activity of the pre- and postsynaptic neurons that the target cell is a favored site for the elaboration of additional synaptic endings (Figure 5.7B). This focusing of synapses occurs despite the presence of numerous other valid target neurons in the immediate vicinity. Therefore, the synaptic terminals made on a target neuron act as sets instead of individual entities during the establishment of neural circuits. This and much other evidence implies that neural activity— action potentials that lead to transmitter release at the synapses in question—is somehow involved in circuit formation. Our initial ideas about competition for limiting amounts of target-derived signaling agents is, in retrospect, only part of a more complex story, as shown by Lichtman's remarkable work on this issue during the last 35 years and counting.

Although these observations were intriguing, it was increasingly clear that understanding what was going on during the formation and maintenance of synapses required a way to directly follow the progress of synaptic contacts on the same nerve or muscle target cell over periods of days, weeks, months, or longer. This goal seemed technically feasible in the peripheral nervous system, and would allow us to watch how competition operated during development, and how synaptic connections continued to be modified in maturity. Because encoding experience during life depends on functional and anatomical changes in neural connectivity, the expectation was that synaptic connections would gradually change over time and that we would be able to witness the process in action. The next step was thus to figure out how to monitor the same synapses chronically.

Our first stab at this was indirect, based on the ability to identify the same neuron in the autonomic ganglia of a living animal on different occasions. Given that each neuronal cell body has a somewhat different appearance in the cobblestone-like pattern of cells visible on the surface of a ganglion (Figure 5.8A), it is not hard to find the same neuron during an initial surgical exposure and at a second such operation after an arbitrarily long interval. We could inject an identified neuron with a nontoxic dye and photograph the configuration of its

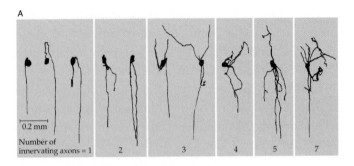

A

B

Figure 5.7   The dependence of neuronal innervation on the geometry of the target cells. A) The proportionality between dendritic complexity and convergence: The more dendrites a neuron has, the greater the number of different nerve cell axons that innervate it, thus affecting the integration of information. B) The synapses arising from a single axon act as a set during innervation. In this example, a single axon leading to the ganglion has been injected with a dye; each cluster represents a group of synapses made on a particular nerve cell in the ganglion (the outline of the ganglion is shown). Despite hundreds of available target neurons, the labeled axon makes many synapses on just a few target cells. (A is after Hume and Purves, 1981; B is from Hume and Purves, 1983)

dendrites. By carrying out the same procedure weeks or months later, we could determine how much, if at all, the dendritic branches changed during the interval. Because the dendrites of ganglion neurons are studded with synapses, any change in the architecture of the dendritic branches would imply ongoing changes in synaptic connectivity. These studies showed that dendrites are slowly being remodeled, and therefore that the synaptic connectivity of the neurons must be slowly changing as well (Figure 5.8B).

A   B

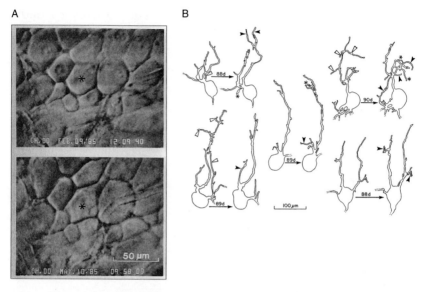

Figure 5.8   Ongoing changes in the synaptic connectivity of nerve cells. A) The appearance of the surface of an autonomic ganglion in the same mouse at an interval of several months; the pattern of cells enables the identification of individual neurons (for example, the one marked with an asterisk) after an arbitrary interval. B) Examples of differences in the configuration of selected portions of the dendrites arising from the same neuron during the interval indicated. Open arrowheads indicate the loss of a dendritic process, and filled ones indicate the addition. (After Purves et al., 1986)

Monitoring the synaptic endings themselves over time would be more revealing, and this is what we set out to do next. The problem in this project was that, unlike the cell bodies in Figure 5.8, synapses are far too small to be directly and routinely injected with a dye. To visualize synapses, we needed a dye that the terminals would quickly absorb when they were bathed in it that would then diffuse away and not damage the endings. Finding such a reagent was a matter of trial and error, and the person who undertook this thankless task was Lorenzo Magrassi, a smart and hard-working medical student from Italy who had come to spend a year in the lab in 1985. Magrassi, who knew a lot about chemistry, applied one plausible reagent after another to synaptic endings on mouse muscles in a dish while he observed the results under a microscope. After many weeks, he finally succeeded in finding a dye that met the criteria. Lichtman (who was by then a faculty member), Magrassi, and I used this approach to watch the same synapses on muscle fibers over months by finding and restaining the same synaptic endings (Figure 5.9). The method also worked for the synaptic endings on identified ganglion cells and synapses on neurons that could be similarly followed over time. In both cases, synaptic terminals gradually changed, slowly on mature muscle fibers and faster on neurons.

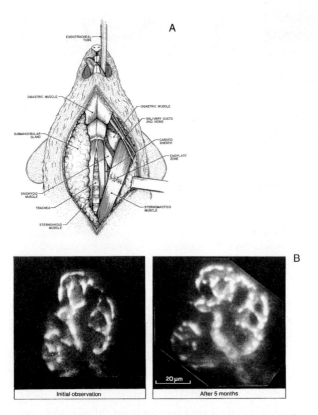

Figure 5.9   Observing synapses directly over time. A) Exposure of an accessible muscle in the neck of an anesthetized mouse. All the synaptic endings on the muscle fibers were then stained by the application of a nontoxic fluorescent dye. B) The same synaptic endings on a single identified muscle fiber (see Figure 5.6A) after an interval of several months, showing small but definite changes. (Reprinted by permission of the publisher from *Body and Brain: A Trophic Theory of Neural Connections* by Dale Purves, p. 111, Cambridge, Mass: Harvard University Press, Copyright © 1988 by the President and Fellows of Harvard College. After Purves, 1986.)

From most perspectives, this effort to understand the formation and maintenance of synapses had been quite successful. But by the mid-1980s, after I had been laboring away on these issues for more than ten years, I didn't think the research was going so well. The reasons were several and were only partly caused by the science. Science proceeds like any other human enterprise, with real and imagined factors determining the mindset of any individual practitioner. With respect to the science, it was not clear at that point what to do next. Directly monitoring synaptic change in muscle fibers and ganglia had been a fine way to start, but no one was going to get very excited about this work if similar studies of synaptic stability in the brain could not extend it. After all, the brain determines behavior and the cognitive processes that I and everyone else ultimately wanted to understand. I had viewed these studies of synapses in ganglia and muscle as simple model systems for understanding what was likely happening in more interesting parts of the nervous system. But the techniques we had been using were difficult enough to apply in the peripheral nervous system, and, for various reasons, they were impossible to apply at that point to synapses in the brain. Within a decade, further advances in molecular biological methods resolved this impasse by providing labels that could be introduced into neurons by gene transfection. This methodology eventually enabled Lichtman and his collaborators and many others to begin tackling these types of problems in the brain, but that possibility was not on the horizon in the mid-1980s.

Other factors were at work as well. Hunt, the head of the department in which I had worked for most of this time, had retired and moved to France. The department's focus had changed to cell biology and, as a result, I (along with Lichtman and Sanes) had moved to the Department of Anatomy and Neurobiology in 1986. Gerry Fischbach was by then running the department because Cowan had left to take a position as the chief scientific administrator at the Howard Hughes Medical Institute. Fishbach is a lovely person and was a fine chairman. But he was a peer, creating a situation quite different from the paternal figure Hunt had been. In the new department, I was among colleagues who were explicitly interested in the brain and had a different outlook on neuroscience than the one I had grown up with at Harvard and University College, and that I had continued to experience in Hunt's department. Finally, my relationship with Lichtman became increasingly awkward. He was by then a faculty member with his own very successful lab, and although we had continued to collaborate, the relationship was no longer the one that I had enjoyed for many years. We were now working on the same issues and, to some degree, had become competitors; the situation was delicate for both of us.

In a couple years, I would be 50, and the undeniable fact of middle age combined with these several circumstances combined to trigger another bout of depression, this one more serious than those I had experienced as a teenager, in college, or in med school. Although I kept coming to work every day, my enthusiasm for what I was doing dwindled. I saw a psychiatrist, who started me on an antidepressant, and when the drug he prescribed didn't work (ironically, it was the one I had worked on in my summer of neuropharmacology research as a med student), the depression deepened. Mainly as a result of my wife's support, the counsel of another psychiatrist, a different medication, and perhaps just the passage of time, I gradually began to see a plausible future again. After all, I still would presumably have more years left in neuroscience at age 50 than the number I had already completed.

Even so, I realized that plugging away on the same issues in the peripheral nervous system would not suffice. Having gradually returned to a better frame of mind, I began to feel that I had achieved enough success to take some bigger scientific chances. I had started out with broad philosophical interests in the brain, but by virtue of my training, the people and the science that I admired, and the overall direction of neuroscience, I was de facto a reductionist. I had worked on some important issues in model systems, but these would never be more than indirectly related to the things that had first interested me about the brain. With perhaps another 20 years or so to go, I felt that I owed it to myself to at least think about doing something that might move beyond the conventional framework that I had assiduously learned, worked within, and taught for the previous two decades. As it had been ever since those inconclusive evening meetings at Harvard where we postdocs tried to come up with a list of challenges for the future, it was difficult to identify a significant problem. And although I was now a senior figure in the field, when I poked my head up and tried to look beyond the boundaries of the mainstream neuroscience I had been practicing, I had no idea what sort of issue to take on next.

6

Exploring brain systems

Despite many years of investment, giving up the work I had been doing in the peripheral nervous system was not that hard. I had gradually come to feel that working at that level would yield diminishing returns, similar to my sense in the 1970s about the future of invertebrate nervous systems as they were then being studied with electrophysiological and anatomical methods. By the mid-1980s, it was hard to ignore the fact that neuroscientists were shifting to what might be described as either a lower or a higher level of interests. The revolution in molecular biology and the powerful methods it provided were attracting many neuroscientists to pursue the organization of the nervous systems of worms, fruit flies, and mice at the molecular genetic level. At the same time, brain imaging techniques based on computer assisted tomography (CAT) and positron emission tomography (PET) were giving birth to a new field that had been dubbed cognitive neuroscience, defined as a wedding of psychology with these and other neuroscientific methods for studying human brain function. If I simply kept going along the path I had been following, I would be stuck somewhere in the middle. Because molecular reductionism had never been appealing to me, the direction I needed to go in seemed inevitably upward.

The growing sense that I should move on to study some aspect of the brain was painfully underscored during a talk I gave at the Cold Spring Harbor Laboratories in the mid-1980s. Three of us neuroscientists had been invited to address the Lab's Winter Meeting, a session attended by the cadre of molecular biologists at the lab that Jim Watson had turned into a major intellectual center since he had taken over as director in 1968. These generally young and frighteningly bright molecular biologists were also looking for new problems to pursue, and they wanted to know the latest developments in neuro-science and how they might tap into them using the new molecular genetic techniques. Chuck Stevens, a synaptic physiologist, spoke first on synaptic noise (the molecular effects of individual transmit-ter molecules interacting with transmitter receptors discussed in Chapter 3). I followed with a talk on the work we had been doing on synaptic development and maintenance in the autonomic nervous system, and David Hubel ended the meeting with a discussion of the work he and Torsten Wiesel had done in the visual system. As Hubel mounted the stage after the end of my presentation, he turned and said pointedly to the audience, "And now for the *real* story!" My face must have reddened as the audience tittered at the gratuitous insult, but his point was not lost on me.

To pursue any sort of study at a "higher" level of the nervous sys-tem, I needed to define a worthwhile problem in the brain, which was not easy. It might seem odd, but after nearly 20 years as a practicing neuroscientist doing what was considered important research and teaching neurobiology to medical students, graduate students, and undergraduates, I actually knew very little about the brain. I had been taught brain neuroanatomy and pathology in medical school, but I soon forgot this information when it proved to be of no great value during my clinical years as a general surgical-house officer, or during the two years I put in as general practitioner in the Peace Corps. The brain's anatomy is enormously complicated and not at all logical. To make matters worse, neurophysiologists (which I was) gen-erally looked down on neuroanatomy as a field, and the work I had been doing since 1973 had provided little reason to delve into issues that concerned the brain. Although I had learned a lot of information by osmosis over the years, absorbing the relevant detail—Figure 6.1 shows only a few elements of brain structure—requires the sort of dedication that comes from the daily demands of clinical practice in neurology or neurosurgery, or a research career specifically directed at one or more brain systems. As a result, I had managed to avoid all but the most superficial knowledge of brain organization, and I would have had trouble answering the questions on the anatomical part of the exam that we routinely gave the first-year medical students.

If I was going to pursue research on some aspect of the brain, it was clear that I would need a good deal of remediation, and the first

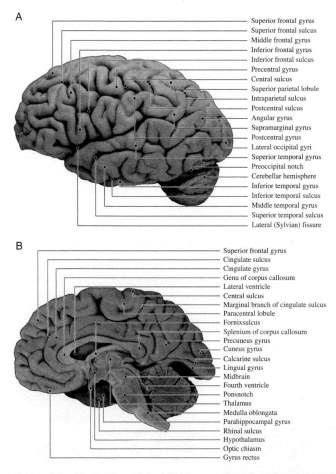

A

- Superior frontal gyrus
- Superior frontal sulcus
- Middle frontal gyrus
- Inferior frontal gyrus
- Inferior frontal sulcus
- Precentral gyrus
- Central sulcus
- Superior parietal lobule
- Intraparietal sulcus
- Postcentral sulcus
- Angular gyrus
- Supramarginal gyrus
- Postcentral gyrus
- Lateral occipital gyri
- Superior temporal gyrus
- Preoccipital notch
- Cerebellar hemisphere
- Inferior temporal gyrus
- Inferior temporal sulcus
- Middle temporal gyrus
- Superior temporal sulcus
- Lateral (Sylvian) fissure

B

- Superior frontal gyrus
- Cingulate sulcus
- Cingulate gyrus
- Genu of corpus callosum
- Lateral ventricle
- Central sulcus
- Marginal branch of cingulate sulcus
- Paracentral lobule
- Fornixsulcus
- Splenium of corpus callosum
- Precuneus gyrus
- Cuneus gyrus
- Calcarine sulcus
- Lingual gyrus
- Midbrain
- Fourth ventricle
- Ponsnotch
- Thalamus
- Medulla oblongata
- Parahippocampal gyrus
- Rhinal sulcus
- Hypothalamus
- Optic chiasm
- Gyrus rectus

Figure 6.1   The major surface features of the human brain in lateral (A) and medial (B) views. (The medial view is of the right cerebral hemisphere as seen from the midline with the left hemisphere removed.) The labeled structures represent only a few of the surface landmarks that guide neuroscientists working on this complex structure; the internal organization of the brain is far more complex. (After Purves, Augustine, et al., 2008)

order of business was to find a good teacher. I was especially lucky in 1987 when Anthony LaMantia, a newly minted neuronanatomist who had just finished his Ph.D. with Pasko Rakic at Yale, got in touch with me about joining the lab as a postdoctoral fellow. Rakic, who had trained with the people who had taught me neuroanatomy and neuropathology at Harvard, was without much question the most accomplished and imaginative neuroanatomist in the country. I had closely followed his work on the development of the primate brain and very much admired what he had done. His knowledge and talent rubbed off on his trainees, and given my new inclination, LaMantia's matriculation in the lab the following year was a godsend. I learned far more from him during the next few years than he learned from me.

Relearning brain anatomy, however, was only preliminary to figuring out a good problem to explore. Typically, investigators extend their research in familiar directions, making an educated guess about an interesting tangent. This is what Anthony and I did, ultimately deciding to tackle a problem in the olfactory system. By 1988, the work on monitoring synapses over time in the peripheral nervous system was winding down for me (although not for Jeff Lichtman, who was just beginning to look directly at synaptic competition on developing muscle fibers in more sophisticated ways). LaMantia and I would have liked to have done something similar in some region of the living brain, but there was no technical way then. So we settled on what we thought was the next best thing: monitoring the development and maintenance of brain modules (Figure 6.2).

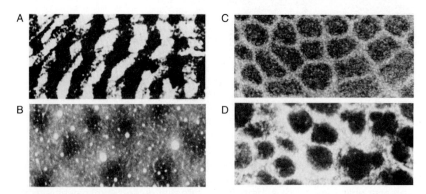

Figure 6.2   Examples of the modular units found in the mammalian brain. A) The striped pattern of ocular dominance columns in the visual cortex of a rhesus monkey. B) The units—whimsically called 'blobs'—in another region of the visual cortex of monkeys. C) The units called 'barrels' in somatic sensory cortex of rats and mice. D) The units called glomeruli in the olfactory bulb of all mammals. In each case, the units are a millimeter or less in width or diameter, and are revealed by staining techniques that enable them to be seen when looking down on the surface of the brain with a low-power microscope. (Purves, D., D. Riddle, and A. LaMantia. "Iterated Patterns of Brain Circuitry (or How the Cortex Gets Its Spots)." Reprinted from *Trends in Neuroscience* 15 (1992): 362–368 with permission from Elsevier.)

Applying the term *modules* to the brain has always been problematic. Many psychologists and mind–brain philosophers use it to refer to the idea that different brain regions are dedicated to particular functions. This concept needs to be carefully qualified because although brain structures and cortical regions are specialized to perform some range of functions, all these structures and regions interact directly or indirectly; even when the interactions are indirect, a surprisingly small number of synaptic linkages connect any one brain region to another. In contrast, neurobiologists use the term *modules* to refer more specifically to small repeating units within specialized brain regions that comprise local groups of functionally related neurons and their connections (Figure 6.2). These units occur in the brains of many mammals, and their prominence had made them a focus of interest and speculation about cortical function since the 1950s. Spanish neuroanatomist Rafael Lorente de Nó, a protégé of Cajál, first noted in the 1920s that many regions of the cerebral cortex comprise iterated elementary units. ("Cortex" refers to the layer of gray matter that covers the cerebral hemisphere, and much of the higher-order neural processing in the brain occurs in cortical circuitry; see Figure 6.1.) This cortical modularity remained largely unexplored until the late 1950s, when electrophysiological experiments indicated an arrangement of repeating units in the brains of cats. Vernon Mountcastle, a leading neurophysiologist who spent his career at Johns Hopkins, reported in 1957 that vertical microelectrode penetrations made in the cortical region of the cat brain that processes mechanical stimulation of the body surface encountered neurons that all responded to the same type of stimulus presented at the same body location. When Mountcastle moved the electrode to nearby locations in this brain area, he again found similar responses of neurons located along a vertical track down through the cortex, although the functional characteristics of the nerve cells were often different from the properties of the neurons along the first track (for example, the nerve cells along one vertical track might respond to touch, and those along another track respond to pressure). A few years after Mountcastle's discovery, Hubel and Wiesel (who were in Kuffler's lab at Johns Hopkins in 1957 and well aware of this work) discovered a similar arrangement in the visual cortex of the cat, and later in the visual cortex of monkeys. These observations, along with evidence in other cortical regions, led Mountcastle and a number of other investigators to speculate that these repeating units represented a fundamental feature of the mammalian cerebral cortex that might be relevant to brain functions that remained poorly understood, including cognitive abilities and even consciousness. The role, if any, of these iterated patterns for brain function remains uncertain. However, the prevalence of these modules and the ongoing interest in their role gave LaMantia and me a way to assess the stability of brain organization monitored over weeks or months.

The modular units we chose to look at first were the glomeruli in the olfactory bulb of the mouse (see Figure 6.2D). These weren't the most interesting or most talked-about cortical patterns—that prize went to the ocular dominance columns in the visual cortex (see Figure 6.2A). However, glomeruli were practical targets to begin with because mice are cheap (a couple of dollars apiece then, compared with several hundred dollars for a cat and closer to a thousand dollars for a monkey); it was clear we would have to use a lot of animals to work out methods for exposing the brain, staining the units with a nontoxic dye, and repeating the procedure to examine the same region weeks later. We succeeded in doing this during the next year or so, asking whether these units in the mouse brain all developed at the same time as the animals grew up, or whether new units were added among the preexisting ones. No one was waiting impatiently for the answers to these questions about the olfactory system, however, and our finding that some units are demonstrably plugged in among the existing ones as mice mature did not elicit much excitement. The focus of interest in cortical modularity was the visual system, and this was part of the brain that had stimulated the most ardent debates about the role of cortical modularity.

As a graduate student in Rakic's lab, Anthony had plenty of experience working with rhesus monkeys, so we turned next to the monkey visual cortex. For a variety of technical reasons, it was impractical to do repeated monitoring in the same monkey as we had done while looking at modules in the olfactory system of the mouse. So we again settled for the next best thing. Given what we had found in the mouse olfactory bulb, it seemed reasonable to look at the overall number of modular units in the visual cortex shortly after birth and in maturity in different monkeys; if the average numbers were significantly different, we could assume that units had been added as monkeys matured, much as we had documented in the olfactory brain of the mouse. The easiest units to look at in the monkey visual cortex were the so-called blobs shown in Figure 6.2B. These units had not attracted the same attention as ocular dominance columns, but they were discrete and could be easily counted (ocular dominance columns form more complex patterns and would have been hard to assess quantitatively; see Figure 6.2A). In 1990, LaMantia and I began working to determine the number of blobs present in the primary visual cortex of newborn monkeys compared to the number in adults.

The project was problematic from the outset. Although I had worked on a lot of different species over the years, including small monkeys, adult rhesus monkeys are large and nasty, and dealing with them is distinctly unpleasant. The expense, the character of the animals, and the knowledge that the project was only a stepping stone to a more direct approach made us hurry along and eventually publish a wrong conclusion. Based on the first few animals in which we counted blobs at birth and in maturity, it seemed reasonably clear that these units were being added. Anxious to stake this claim in the monkey visual system and convinced that the results we had seen in the mouse olfactory bulb indicated a general rule, we went ahead and published a short paper to that effect. However, when we completed the study with a larger complement of monkeys, there was no statistically significant difference in the initial and mature number of blobs in the visual cortex. We corrected our mistake in the full report of the project, and no one seemed to have paid much attention to our error, but I realized that I had pushed too hard in the interests of being recognized as a player in my newly adopted field and the community of brain scientists. This minor fiasco left me considerably less confident about making the transition from the peripheral nervous system to the brain. It also left me with the need to determine another research direction. To judge from our work on blobs and other evidence we should have paid more attention to, modular units in the visual cortex seemed pretty stable.

While this was going on, my scientific and personal life changed in another important way. In 1990, I had accepted an offer from Duke University to start up a Department of Neurobiology there, and Shannon and I and our younger daughter had moved to North Carolina (which is where LaMantia and I carried out the work on monkey blobs). Most universities wanted to follow the example Kuffler had set at Harvard in 1966 by forming departments of neurobiology; many had done so as neuroscience, and the funding for it, grew rapidly in the 1970s and 1980s. But starting up a new department meant committing a great deal of money and overcoming the political opposition of existing departments that did not want to give up any of their own resources or clout. As a result, most places had made the sort of compromise Max Cowan had engineered at Washington University, marrying neurobiology to an existing department of anatomy or physiology. However, Duke had raised the large sum required to hire and set up about a dozen new faculty and had built a new building to house the department. This largesse coupled with the quality of the university and its ambitions presented an opportunity that was hard to turn down, even though it promised some administrative work that I had always shunned.

The move turned out to be significant in many ways, almost all of them positive. My less-than-robust mental state when I accepted the job at Duke benefited greatly from the change of scene and the challenge of starting up the new department. I had been at Washington University nearly 17 years, and my crankiness at the end of that time, the problematic relationship with Lichtman, and my desire for a new scientific start were all resolved in one fell swoop. The move was also a big plus for Shannon. Shannon (known as Shannon Ravenel in the publishing world) had been a successful young editor with Houghton Mifflin in Boston when we married in 1968, but she had given up her job when we moved to London in 1971. During our first few years in St. Louis, she had taken on a series of minor editorial jobs to make ends meet that were as demeaning to her as it would have been for me to teach junior high health science at that point in my career. Her professional situation had improved in 1977 when Houghton Mifflin asked her to become the series editor of their annual anthology *Best American Short Stories*, a job she could do in St. Louis that entailed reading and selecting stories from all the short fiction published in American and Canadian periodicals each year. Shannon's situation changed for the better again in 1982 when her friend and mentor Louis Rubin asked her if she would be interested in starting a literary publishing company in Chapel Hill, where he was then professor of English at the University of North Carolina. She agreed and had been exercising her role in this new venture from St. Louis. But as Algonquin Books of Chapel Hill grew more successful, this arrangement had become increasingly awkward. My move to Duke (only 10 miles from Chapel Hill) solved problems for both of us. (Workman Publishing Co. in New York purchased Algonquin Books in 1989, and it continues to flourish in Chapel Hill.) The administration at Washington University made no particular effort to keep me there (whether because they realized I was going to go anyway or because they didn't really care), and we arrived in North Carolina in summer 1990.

The first of several postdoctoral fellows to join my new lab at Duke, in addition to LaMantia, was David Riddle, a recent Ph.D. from the University of Michigan. The new direction that seemed most attractive involved yet another region of the brain that had long been of interest: the somatic sensory system that Sherrington, Mountcastle, and many other others had studied. This region of the sensory brain is responsible for processing the mechanical information arising from the variety of receptor organs in skin and subcutaneous tissues that initiate the sensations we experience as touch, pressure, and vibration (pain and temperature sensations entail a related but largely separate system). Although not as thoroughly plowed as the visual system (and intrinsically less interesting to most people), the somatic sensory system had some advantages for us. The major attraction was visibility of the animal's body surface in the relevant part of the cortex (see Figures 6.2C and 6.3). Although this "map" of the body surface is apparent in the brains of many species, including primates, it is especially easy to see in rats and mice. Tom Woolsey, one of my colleagues in the Department of Anatomy and Neurobiology at Washington University, discovered the visible body map in rodents in 1970 when he was working with Hendrik van der Loos at Johns Hopkins, shortly after graduating from medical school there. As a result of their shape, these units were called barrels (only the top of the barrel is apparent in Figures 6.2C and 6.3A) and had been much studied ever since. Each barrel in the cortex corresponds to a richly innervated peripheral sensory structure, such as a particular whisker on the animal's face or a digital pad on one of the paws. Barrels are thus processing centers for sensory information arising from the corresponding peripheral structure, and can be made visible because of their higher rate of neural activity relative to the surrounding cortex. The higher level of neural activity in barrels means a higher rate of cellular metabolism, which causes barrels to preferentially take up reagents such as mitochondrial enzymes stains that reflect nerve cell metabolism.

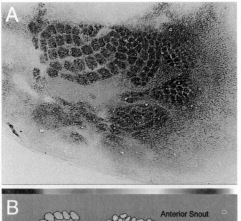

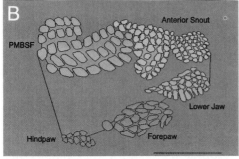

Figure 6.3   The somatic sensory cortex in the rat brain, made visible by activity-dependent enzyme staining. A) The map of the rat's body that is apparent when looking down on the flattened surface of this region of the animal's cerebral cortex (a similar map is present in each cerebral hemisphere). Each of the darkly stained elements (the barrels) corresponds to a densely innervated peripheral structure such as a facial whisker or one of the pads on the forepaws or hind paws. B) Tracing of the map in (A), indicating the corresponding body parts; the color-coding shows the relative level of neural (metabolic) activity in each module, with yellow signifying higher activity. The red bar indicates 2 millimeters. (From Riddle, et al., 1993)

Riddle and I (and, eventually, Gabriel Gutierrez) wanted to see if the regions of sensory cortex that experienced more neural activity during maturation captured more cortical area than less active brain regions. We could explore this question by measuring the area occupied by different components of the somatic sensory map at different ages, asking whether the more metabolically active areas grew faster (see Figure 6.3B). If we could establish this correlation, it would imply that the neural activity generated by an animal's experience in life was being translated into the allocation of more cortical area for processing the relevant information (showing that neural activity influences cortical connectivity and the amount of cortex devoted to specific tasks). This turned out to be the case: The more active cortical regions expanded relatively more during maturation than less active ones. But having established this point, it was not clear how to go forward in a direction that would warrant a further effort along these lines.

By the early 1990s, I had spent about five years working on these various projects in the brain, felt more knowledgeable about the issues, and had finally acquired a reasonable working knowledge of brain anatomy. To show my willingness to participate in the grunt work of the new department, I was teaching the Duke med students neuroanatomy each spring. And although I was certainly no expert, I no longer embarrassed myself when answering questions from the students and could easily have passed the first-year exam that I had imagined failing unceremoniously a few years before. Work on the growth and organization of the cortex as a function of activity continued, and as much from scientific boredom as from any clear goal, I began thinking about visual perception, fiddling around with some small projects that seemed interesting but minor asides to the mainstream neuroscience that was plodding along in the lab.

Perception—what we actually see, hear, feel, smell, or taste—is generally thought of as the end product of sensory processing, eventually leading to appropriate motor or other behavior. But perception is far more complicated than this garden-variety interpretation. For one thing, we are quite unaware of the consequences of most of the sensory processing that goes on in our brains. For example, think of the sensory information the brain must process to keep your car on the road when you're driving along and focusing on other concerns. We obviously see perfectly well when our minds are otherwise engaged, but we are not aware of seeing in the usual sense. Why, then, do we ever need to be aware of sensory information? Furthermore, what we see or otherwise experience through the senses rarely tallies with corresponding physical measurements. Why are these discrepancies so rampant? And what do these discrepancies have to do with the longstanding philosophical inquiry into the question of how we can know the world through our senses in the first place? After I had begun to work on sensory systems, these and other questions about perception kept intruding and were growing harder to ignore. They are, after all, pretty interesting.

Similar to most mainstream neuroscientists, I was leery of devoting much time to questions that are generally looked down on as belonging to psychology or, worse yet, philosophy. I had first gotten a sense of this bias as a postdoc in Nicholls's lab circa 1970, when we had lunch every day with the Hubel and Wiesel lab. Because they were working on vision, Hubel and Wiesel were familiar with many of the controversies and issues in visual perception. The only psychologists I remember them taking seriously, however, were people such as Leo Hurvich and Dorothea Jameson, who devoted their careers to painstaking psychophysical documentation of people's perceptions of brightness and color, and models of how these phenomena might be explained. When less rigorous psychologists came up in conversation, Hubel would refer to them as "chuckleheads," a term he used often (and did not limit to psychologists). Likewise, the rest of my mentors and colleagues at Harvard, University College, and later Washington University didn't waste much thought on psychology or its practitioners; this sort of work was deemed irrelevant to the rapid progress of the reductionist neuroscience that nearly all of us were doing. Psychology as science was considered not up to par, and philosophical questions were simply nonstarters.

Some basis exists for this general lack of enthusiasm. Even though I have directed a center for cognitive neuroscience for the last six years ("cognitive neuroscience" being a more fashionable name for much of the territory covered by modern psychology), many psychologists do seem to be a bit chuckleheaded. This failing is not through any deficiency of native intelligence, but arises from the difficulty we all have in transcending the tradition in which we were trained. In science, as with anything else, we tend to run with the pack. For decades, psychologists had been mired in gestalt or behaviorist theories and their residua, and had not run a very good race. When physicists and chemists referred to biologists before sufficient cellular and molecular information hardened the field during the twentieth century, they no doubt lodged the same complaint about their relative "soft-headedness." These biases notwithstanding, perception and the psychological and philosophical issues involved were intellectual quicksand: Fascination with the relationship between the real world and what we end up perceiving was getting me more and more deeply stuck.

I undertook the first of several miniprojects on perceptual issues in 1994 with another postdoc, Len White, and I think we both regarded it simply as an amusing diversion from the especially tedious project we were conducting. Len was a superbly trained neuroanatomist who had recently earned his doctorate with Joel Price, one of Cowan's original hires at Washington University. We had been looking at the neural basis of right- and left-handedness by laboriously measuring the cortical hand region in the two hemispheres of human brains. Based on the effect of activity on the allocation of brain space that Riddle, Gutierrez, and I had seen in rodents, we thought that human right-handers would very likely have more cortex devoted to that hand in the left hemisphere, where the right hand is represented, and conversely for left-handers. Thus, White and I were in the process of measuring this region in hundreds of microscopical sections taken from brains that had been removed during autopsy.

People are not just right- or left-handed—they are also right- or left-footed and, interestingly, right- or left-eyed. To leaven the load of measuring the right- and left-hand regions in what ended up being more than 60 human brains, we started thinking about right- and left-eyedness. The minimal question we asked about perception was whether people who were either right-eyed or left-eyed when sighting with one eye (such as aiming a rifle) routinely expressed this preference in viewing the world with both eyes. We covered a large panoramic window in the Duke Neurobiology building with black paper into which we had cut about a hundred holes the diameter of a tennis ball. We then asked subjects to simply wander around the room and look at the scene outside through the holes, which they would necessarily have to do using one eye or the other. This setup mimicked the everyday situation in which we look at the objects in a scene that lies beyond occluding objects in the foreground (if you look around the room in which you are reading this, there will be many examples). As subjects looked at the outside world through the holes from a meter or two away, we monitored whether they used the right or left eye to do so, and whether the eye they used matched the eyedness they showed in a standard monocular sighting task. It did match, although as far as I know, no one paid attention to the short paper that we published on the topic. However, doing this work and thinking about the issues involved was more fun than measuring the hand region in human brains (which, as it turned out, showed no significant difference between the cortical space devoted to the right and left hand in humans).

One thing leads to another in science, and our little eye project got everyone in the lab interested in perception, at least to some degree. It also raised eyebrows among my colleagues in the Department of Neurobiology. When they walked by and saw the papered-over window with people wandering around looking out through little holes, it was apparent that some weird things were beginning to occur in my lab. The faculty I had recruited to the department seemed either mildly bewildered or amused at the apparent flakiness of what we were doing, which was very far from neurobiology as they understood it. Instead of leading the troops into battle, the general was apparently playing tiddlywinks.

Another project in perception that we undertook at about the same time was just as peculiar but less trivial, and accelerated my transition (colleagues might have thought "downward slide") toward a focus on perception. Another postdoc, Tim Andrews, had received his degree in the United Kingdom working on trophic interactions and neural development, and had come expecting to work with me on some related issue in the central nervous system. But when he arrived, the quicksand phenomenon affected him as well, and he became the first postdoc to work primarily on perception. (Andrews continues to work on visual perception as a member of the psychology faculty of York University in the United Kingdom.) The eyedness project uncovered a lot of interesting literature that I had never come across, including several papers by Charles Sherrington describing some little-known experiments on vision that he carried out in the early 1900s. As already mentioned, Sherrington was one of the pioneers of modern neurophysiology and the mentor of John Eccles, who was, in turn, the mentor of Kuffler and Katz. The main body of Sherrington's highly influential work was on motor reflexes, and one of his principal findings was that actions were always routed through a "final common pathway." This meant that the output of all the neural processing that goes on in the motor regions of the brain ultimately converges onto the spinal motor neurons that innervate skeletal muscles, which generate motor behavior (see Figure 1.3A). It was therefore natural for him to ask whether a similar principle might apply to sensory processing. Might all sensory processing in the brain likewise be funneled into a final common pathway, which would lead to perception in a given modality such as vision or audition?

Sherrington recognized that the sensory nervous system provided at least one good venue for addressing this question, namely the processing carried out by the neurons in the visual system that are related, respectively, to the right and left eyes. He also knew, as we all do from subjective experience, that the scenes we end up seeing are seamless combinations of the information the left and right eyes process individually, a combination that is experienced as if we were viewing the world with a single eye in the middle of the face (this subjective sense is referred to as cyclopean vision). Because the line of sight of each eye is directed at nearby objects from a different angle, the left and right eyes generate quite different images of the world. To convince yourself of this, simply hold up any three dimensional object at reading distance and compare what you see looking at it with one eye and then the other; the left-eye view enables you to see parts of the left side of the object that the right-eye view does not include, and the right-eye view enables you to see parts on the right side that are hidden from the left eye. How these two views are integrated was an especially challenging question for Sherrington and remains just as challenging today.

Sherrington used the initially separate processing of visual information by each eye to test his idea of a final common sensory pathway by taking advantage of the perceptual phenomenon called 'flicker-fusion'. The flicker-fusion frequency is the rate at which a light needs to be turned on and off before the off periods are no longer noticed and the light appears to be always on. The cycling rate at which this fusion happens in humans is about 60 times a second, and is important in lighting, movies, and video. The room lights that we see as being "on" are actually going on and off 120 times a second as a result of the alternating current used in the U.S. power grid, and movies in early decades of the twentieth century flickered because the stills were presented at less than the flicker-fusion frequency. Sherrington surmised that if there were a final common pathway for vision, then the flicker-fusion frequency when the two eyes are synchronously stimulated ought to be different from the frequency observed when the two eyes are stimulated alternately (meaning one eye experiencing light when the other eye experienced dark; Figure 6.4). If the information from the two eyes is brought together in a final common pathway in the visual brain, the combined asynchronous left and right eye stimulation should be perceived as continuous light at roughly half the normal flicker-fusion frequency (see Figure 6.4B). However, Sherrington found that the on-off rate at which a flashing light becomes steady is virtually identical in the two circumstances. Based on this observation (which Andrews and I confirmed), Sherrington concluded rather despairingly that the two retinal images must be processed independently in the brain and that their union in perception must be "psychical" instead of physiological, thus lying beyond his ability (or interest) to pursue. Given this outcome, Sherrington returned to studies of the motor system and never worked on vision again.

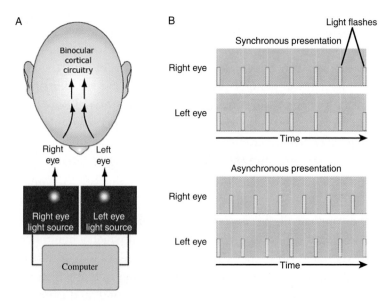

Figure 6.4   A modern version of Sherrington's experiment, in which the perception elicited by synchronous and asynchronous stimulation of the two eyes with flashes of light is compared. A) A computer triggers independent light sources whose relationship can be precisely controlled; one series of flashes is seen by only the left eye, and the other series by only the right. B) Diagram illustrating synchronous versus asynchronous stimulation of the left and right eyes. The fact that observers report the same flicker-fusion frequency whether the two eyes are stimulated synchronously or asynchronously presents a deep perceptual puzzle. (After Andrews et al., 1996)

For better or for worse, Andrews and I and other students and fellows in the lab continued down this path during the next couple of years, carrying out a series of projects on visual perception that examined other odd phenomena, such as the wagon wheel illusion in continuous light, the rate of disappearance of the images generated by retinal blood vessels, the strange way we perceive a rotating wire-frame cube, and the rivalry between percepts that occurs when one stares long enough at a pattern of vertical and horizontal stripes. All this time, the lab was conducting conventional projects on handedness, on the way cortical organization was affected by the prevalence of differently oriented contours in natural scenes, and on other unfinished business in mainstream neurobiology. The reality, however, was an ever-greater interest in perception and less devotion to issues of brain structure and function using the sorts of electrophysiological and anatomical tools I was familiar with.

The tipping point came in 1996. By then, I was in my late 50s, and after four or five years puttering around with how activity affects brain organization, I hadn't stumbled across anything that was deeply exciting. With perhaps 10 or 15 good working years left, I began to think that I should spend all my remaining time working on perception. I had learned enough about the brain and the visual system to have a strong sense that the attempt to explain perception in terms of a logical hierarchy of neuronal processing, in which properties of visual neurons at one stage determine those at the next, was stuck in some fundamental way. Hubel and Wiesel had set about reaching this goal shortly before I met them as a med student in 1961, and we budding neuroscientists in the early 1970s thought it would soon be reached. But despite an effort spanning 40 years in which the properties of many types of nerve cells in the visual system had been carefully documented, the goal did not seem much closer. No percept had been convincingly explained in these terms, a wealth of visual perceptual phenomena remained mysterious, and how the orderly structure of the visual system is related to function remained unclear. When that much time goes by in science without much success, it usually means that the way people are thinking about a problem is on the wrong track.

I was pretty sure that, given my age, the raised eyebrows of my colleagues, and my lack of any serious credentials in vision science, it would be an uphill fight to support a lab focused explicitly on perception. (Up to that point, I had been using money from grants for the conventional work in the lab to support our forays into perception.) I also sensed that I was starting to be seen as something of an oddball, and I was less often being invited to speak at high-profile meetings or asked to lecture at other institutions. On the other hand, work on perception doesn't cost much to do, and I knew that if I didn't take the plunge at that point, there would be no second chance. And so I plunged.

7

The visual system:
Hubel and Wiesel redux

I don't think many neuroscientists would dispute the statement that the work David Hubel and Torsten Wiesel began in the late 1950s and continued for the next 25 years provided the greatest single influence on the ways neuroscientists thought about and prosecuted studies of the brain during much of the second half of the twentieth century. Certainly, what they were doing had never been very far from my own thinking, even while working on the formation and maintenance of synaptic connections in the peripheral nervous system. To explain the impact of their work and to set the stage for understanding the issues discussed in the remaining chapters, I need to fill in more information about the visual system, what Hubel and Wiesel actually did, and how they interpreted it.

Presumably because we humans depend so heavily on vision, this sensory modality has for centuries been a focus of interest for natural philosophers and, in the modern era, neuroscientists and psychologists. By the time Hubel and Wiesel got into the game in the 1950s, a great deal was already known about the anatomy of the system and about the way light interacts with receptor cells in the retina to initiate the action potentials that travel centrally from retina to cortex, ultimately leading to what we see. The so-called primary visual pathway (Figure 7.1) begins with the two types of retinal receptors, rods and cones, and their transduction of light energy.

The visual processing that rods initiate is primarily concerned with seeing at very low light levels, whereas cones respond only to greater light intensities and are responsible for the detail and color qualities that we normally think of as defining visual perception.

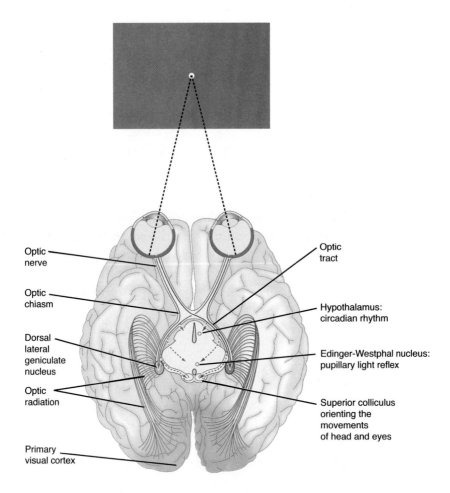

Figure 7.1   The primary visual pathway carries information from the eye to the regions of the brain that determine what we see. The pathway entails the retinas, optic nerves, optic tracts, dorsal lateral geniculate nuclei in the thalamus, optic radiations, and primary (or striate) and adjacent secondary (or extrastriate) visual cortices in each occipital lobe at the back of the brain (see Figures 7.2 and 7.3). Other central pathways to targets in the brainstem (dotted lines) determine pupil diameter as a function of retinal light levels, organize and motivate eye movements, and influence circadian rhythms. (After Purves and Lotto, 2003)

However, the primary visual pathway is anything but simple. Following the extensive neural processing that takes place among the five basic cell classes found in the retina, information arising from both rods and cones converges onto the retinal ganglion cells, the neurons whose axons leave the retina in the optic nerve. The major targets of the retinal ganglion cells are the neurons in the dorsal lateral geniculate nucleus of the thalamus, which project to the primary visual cortex (usually referred to as V1 or the striate cortex) (Figure 7.2).

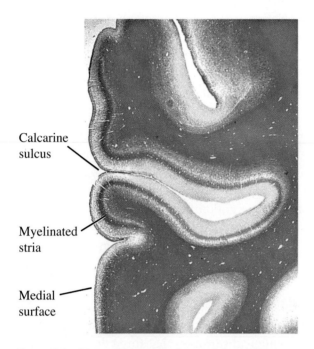

Calcarine
sulcus

Myelinated
stria

Medial
surface

Figure 7.2 Photomicrograph of a section of the human primary visual cortex, taken in the plane of the face (see Figure 7.1). The characteristic myelinated band, or stria, is why this region of cortex is referred to as the striate cortex (myelin is a fatty material that invests most axons in the brain and so stains darkly with reagents that dissolve in fat, such as the one used here). The primary visual cortex occupies about 25 square centimeters (about a third of the surface area of a dollar bill) in each cerebral hemisphere; the overall area of the cortical surface for the two hemispheres together is about 0.8 square meters (or, as my colleague Len White likes to tell students, about the area of a medium pizza). Most of the primary visual cortex lies within a fissure on the medial surface of the occipital lobe called the calcarine sulcus, which is also shown in Figure 6.1B. The extrastiate cortex that carries out further processing of visual information is immediately adjacent (see Figure 7.3). (Courtesy of T. Andrews and D. Purves)

Although the primary visual cortex (V1) is the nominal terminus of this pathway, many of the neurons there project to additional areas in the occipital, parietal, and temporal lobes (Figure 7.3). Neurons in

V1 also interact extensively with each other and send information back to the thalamus, where much processing occurs that remains poorly understand. Because of the increasing integration of information from other brain regions in the visual cortical regions adjacent to V1, these higher-order cortical processing regions (V2, V3, and so on) are called visual association areas. Taken together, they are also referred to as extrastriate visual cortical areas because they lack the anatomically distinct layer that creates the striped appearance of V1 (see Figure 7.2). In most conceptions of vision, perception is thought to occur in these higher-order visual areas adjacent to V1 (although note that what *occur* means in this statement is not straightforward).

By the 1950s, much had also been learned about visual perception. The seminal figures in this aspect of the history of vision science were nineteenth-century German physicist and physiologist Hermann von Helmholtz, and Wilhelm Wundt and Gustav Fechner, who initiated the modern study of perception from a psychological perspective at about the same time. However, Helmholtz gave impetus to the effort to understand perception in terms of visual system physiology, and his work was the forerunner of the program Hubel and Wiesel undertook nearly a century later.

A good example of Helmholtz's approach is his work on color vision. At the beginning of the nineteenth century, British natural philosopher Thomas Young had surmised that three distinct types of receptors in the human retina generate the perception of color. Although Young knew nothing about cones or the pigments in them that underlie light absorption, he nevertheless contended in lectures he gave to the Royal Society in 1802 that three different classes of receptive "particles" must exist. Young's argument was based on what humans perceive when lights of different wavelengths (loosely speaking, lights of different colors) are mixed, a methodology that had been used since Isaac Newton's discovery a hundred years earlier that light comprises a range of wavelengths. Young's key observation was that most color sensations can be produced by mixing appropriate amounts of lights from the long-, middle-, and short-wavelength regions of the visible light spectrum (mixing lights is called color addition and is different from mixing pigments, which subtracts particular wavelengths from the stimulus that reaches the eye by absorbing them).

(A) Lateral

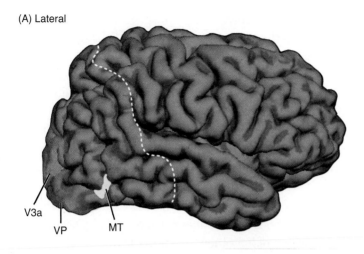

(B) Medial

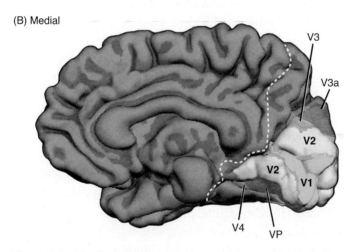

Figure 7.3 The higher-order visual cortical areas adjacent to the primary visual cortex, shown here in lateral (A) and medial (B) views of the brain. The primary visual cortex (V1) is indicated in green; the additional colored areas with their numbered names are together called the extrastriate or visual association areas and occupy much of the rest of the occipital lobe at the back of the brain (its anterior border is indicated by the dotted line). (After Purves and Lotto, 2003)

Young's theory was largely ignored until the latter part of the nineteenth century, when it was revived and greatly extended by Helmholtz and James Clerk Maxwell, another highly accomplished physicist interested in vision. The ultimately correct idea that humans have three types of cones with sensitivities (absorption spectra) that peak in the long, middle, and short wavelength ranges, respectively, is referred to as trichromacy, denoting the fact that most human color sensations can be elicited in normal observers by adjusting the relative activation of the three cone types (see Chapter 9). The further hypothesis that the relative activation explains the colors we actually see is called the trichromacy theory, and Helmholtz spotlighted this approach to explaining perception. Helmholtz's interpretation was that perceptions (color perceptions, in this instance) are a direct consequence of the way receptors and the higher-order neurons related to them analyze and ultimately represent stimulus features and, therefore, the features of objects in the world. For Helmholtz and many others since that era, the feature that color perceptions represent is the nature of object surfaces conveyed by the spectrum of light they reflect to the eye.

This mindset that sensory systems represent the features of objects in the world was certainly the way I had supposed the sensory components of the brain to be working—and as far as I could tell, it was how pretty much everyone else thought about these issues in the 1960s and 1970s. By the same token, I took exploring the underlying neural circuitry (the work Hubel and Wiesel were undertaking) to be the obvious way to solve the problem of how the visual system generates what we see. The step remaining was the hard work needed to determine how the physiology of individual visual neurons and their connections in the various stations of the visual pathway were accomplishing this feat.

Using the extracellular recording method they had developed in Kuffler's lab at Johns Hopkins, Hubel and Wiesel were working their way up the primary visual pathway in cats and, later, in monkeys. At each stage in the pathway—the thalamus, primary visual cortex, and, ultimately, extratstriate cortical areas (see Figures 7.1–7.3)—they carefully studied the response characteristics of individual neurons in the type of setup that Figure 7.4 illustrates, describing the results in terms of what are called the receptive field properties of visual neurons. Their initial studies of neurons in the lateral geniculate nucleus of the thalamus showed responses that were similar to the responses of the retinal output neurons (retinal ganglion cells) that Kuffler had described. Despite this similarity, the information the axons carried from the thalamus to the cortex was not exactly the same as the information coming into the nucleus from the retina, indicating some processing by the thalamus. The major advances, however, came during the next few years as they studied the responses of nerve cells in the primary visual cortex. The key finding was that, unlike the relatively nondescript responses to light stimuli of visual neurons in the retina or the thalamus, cortical neurons showed far more varied and specific responses. On the surface, the nature of these responses seemed closely related to the features we end up seeing. For example, the rather typical V1 neuron illustrated in Figure 7.4 responds to light stimuli presented at only one relatively small locus on the screen (defining the spatial limits of the neuron's receptive field), and only to bars of light. In contrast, neurons in the retina or thalamus respond to any configuration of light that falls within their receptive field. Moreover, many V1 neurons are selective for orientation and direction of movement, responding vigorously to bars only at or near a particular angle on the screen and moving in a particular direction. These receptive field properties were the beginning of what has eventually become a long list, including selective responses to the lengths of lines, different colors, input from one eye or the other, and the different depths indicated by the somewhat different views of the two eyes. Based on this rapidly accumulating evidence, it seemed clear that visual cortical neurons were indeed encoding the features of retinal images and, therefore, the properties of objects in the world.

Important as these observations were, amassing this foundational body of information about the response properties of visual neurons was not Hubel and Wiesel's only contribution. At each stage of their investigations, they used imaginative and often new anatomical methods to explore the organization of the thalamus, the primary visual cortex, and some of the higher-order visual processing regions. They also made basic contributions to understanding cortical development as they went along, work that might eventually stand as their greatest legacy. Hubel and Wiesel knew from the studies just described that neurons in V1 are normally innervated by thalamic inputs that can be activated by stimulating the right eye, the left eye, or both eyes (Figure 7.5). What would happen to the neural connections in the cortex if one eye of an experimental animal was closed during early development, depriving the animal of normal visual experience through that eye? Although most of the neurons in V1 are activated to some degree by both eyes (Figure 7.5A), when they closed one eye of a kitten early in life and studied the brain after the animal had matured

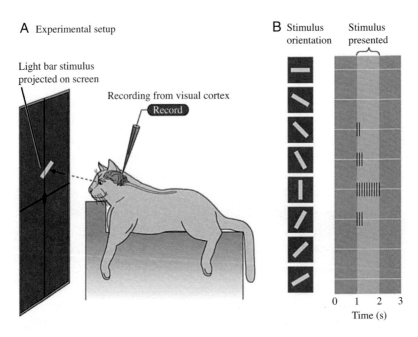

Figure 7.4   Assessing the responses of individual neurons to visual stimuli in experimental animals (although the animal is anesthetized, the visual system continues to operate much as it would if the animal were awake). A) Diagram of the experimental setup showing an extracellular electrode recording from a neuron in the primary visual cortex of a cat (which is more anterior in the brain than in humans). By monitoring the responses of the neuron to stimuli shown on a screen, Hubel and Wiesel could get a good idea of what particular visual neurons normally do. B) In this example, the neuron being recorded from in V1 responds selectively to bars of light presented on the screen in different orientations; the cell fires action potentials (indicated by the vertical lines) only when the bar is at a certain location on the screen and in a certain orientation. These selective responses to stimuli define each neuron's receptive field properties. (After Purves, Augustine, et al., 2008)

(which takes about six months in cats), they found a remarkable change. Electrophysiological recordings showed that very few neurons could be driven from the deprived eye: Most of the cortical cells were now being driven by the eye that had remained open (Figure 7.5B). Moreover, the cats were behaviorally blind to stimuli presented to the deprived eye, a deficit that did not resolve even if the deprived eye was subsequently left open for months. The same manipulation in an adult cat—closing one eye for a long period—had no effect on the responses of the visual neurons. Even when they closed one eye for a year or more, the distribution of V1 neurons driven by one eye and the animals' visual behavior tested through the reopened eye were indistinguishable from normal (Figure 7.5C). Therefore, between the time a kitten's eyes open (about a week after birth) and a year of age, visual experience determines how the visual cortex is wired, and does so in a way that later experience does not readily reverse.

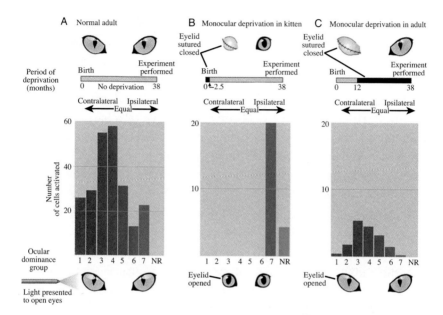

Figure 7.5 The effect on cortical neurons of closing one eye in a kitten. A) The distribution observed in the primary visual cortex of normal adult cats by stimulating one eye or the other. Cells in group 1 are activated exclusively by one eye (referred to here as the contralateral eye), and cells in group 7 are activated exclusively by the other (ipsilateral) eye. Neurons in the other groups are activated to varying degrees by both eyes (NR indicates neurons that could not be activated by either eye). B) Following closure of one eye from one week after birth until about two and a half months of age, no cells could be activated by the deprived (contralateral) eye. C) In contrast, a much longer period of monocular deprivation in an adult cat (from 12 to 38 months of age in this example) had little effect on ocular dominance. (After Purves, Augustine, et al., 2008)

The clinical, educational, and social implications of these results are hard to miss. In terms of clinical ophthalmology, early deprivation in developed countries is most often the result of strabismus, a misalignment of the two eyes caused by deficient control of the direction of gaze by the muscles that move the eye. This problem affects about 5% of children. Because the resulting misalignment produces double vision, the response of the visual system in severely afflicted children is to suppress the input from one eye (it's unclear exactly how this happens). This effect can eventually render children blind in the suppressed eye if they are not treated promptly by intermittently patching the good eye or intervening surgically to realign the eyes. A prevalent cause of visual deprivation in children in underdeveloped countries is a cataract (opacification of the lens) caused by diseases such as river blindness (an infection caused by a parasitic worm) or trachoma (an infection caused by a small, bacteria-like organism). A cataract in one eye is functionally equivalent to monocular deprivation in experimental animals, and this defect also results in an irreversible loss of visual acuity in the untreated child's deprived eye, even if the cataract is later removed. Hubel and Wiesel's observations provided a basis for understanding all this. In keeping with their findings in experimental animals, it was also well known that individuals deprived of vision as adults, such as by accidental corneal scarring, retain the ability to see when treated by corneal transplantation, even if treatment is delayed for decades.

The broader significance of this work for brain function is also readily apparent. If the visual system is a reasonable guide to the development of the rest of the brain, then innate mechanisms establish the initial wiring of neural systems, but normal experience is needed to preserve, augment, and adjust the neural connectivity present at birth. In the case of abnormal experience, such as monocular deprivation, the mechanisms that enable the normal maturation of connectivity are thwarted, resulting in anatomical and, ultimately, behavioral changes that become increasingly hard to reverse as animals grow older. This gradually diminishing cortical plasticity as we or other animals mature provides a neurobiological basis for the familiar observation that we learn anything (language, music, athletic skills, cultural norms) much better as children than as adults, and that behavior is much more susceptible to normal or pathological modification early in development than later. The implications of these further insights for early education, for learning and remediation at later stages of life, and for the legal policies are self-evident.

Hubel and Wiesel's extraordinary success (Figure 7.6) was no doubt the result of several factors. First, as they were always quick to say, they were lucky enough to have come together as fellows in Kuffler's lab shortly after he had determined the receptive field properties of neurons in the cat retina—the approach that, with Kuffler's encouragement, they pursued as Kuffler followed other interests (an act of generosity not often seen when mentors latch on to something important). Second, they were aware of and dedicated to the importance of what they were doing; the experiments were difficult and often ran late into the night, requiring an uncommon work ethic that their medical training helped provide. Finally, they respected and complemented each other as equal partners. Hubel was the more eccentric of the two, and I always found him somewhat daunting. He had been an honors student in math and physics at McGill, and whether solving the Rubik's cube that was always lying around the lunchroom or learning how to program the seemingly incomprehensible PDP 11 computer that he had purchased for the lab, he liked puzzles and logical challenges. He asked tough and highly original questions in seminars or lunchroom conversations and made everyone a little uneasy by taking snapshots with a miniature camera about the size of a cigarette lighter that he carried around. He was hard to talk to when I sought him out for advice as a postdoc, and I couldn't help feeling that his characterization of lesser lights as "chuckleheads" was probably being applied to me. These quirks aside, he is the neuroscientist I have most admired over the years.

Although Wiesel shared Hubel's high intelligence and dedication to the work they were doing, he was otherwise quite different. Open and friendly with everyone, he had all the characteristics of the natural leader of any collective enterprise. Torsten became the chair of the Department of Neurobiology at Harvard when Kuffler stepped down in 1973 and, after moving to Rockefeller University in 1983, was eventually appointed president there, a post he served in with great success from 1992 until his retirement in 1998 at the age of 74. In contrast, Hubel had been appointed chair of the Department of Physiology at Harvard in 1967, but he quit after only a few months and returned to the Department of Neurobiology when he apparently discovered that he did not want to handle all the problems that being a chair entails. (Other reasons might have contributed, based on the response of the Department of Physiology faculty to his managerial style, but if so, I never heard them discussed.)

This brief summary of what Hubel and Wiesel achieved gives some idea of why their influence on the trajectory of "systems-level" neuroscience in the latter decades of the twentieth century was so great. The wealth of evidence they amassed seemed to confirm

Figure 7.6   David Hubel and Torsten Wiesel talking to reporters in 1981, when they were awarded that year's Nobel Prize in Physiology or Medicine. (From Purves and Litchman, 1985)

Helmholtz's idea that perceptions are the result of the activity of neurons that effectively detect and, in some sense, report and represent in the brain the various features of retinal images. This strategy seems eminently logical; any sensible engineer would presumably want to make what we see correspond to the real-world features of the objects that we and other animals must respond to with visually guided behavior. This was the concept of vision that I took away from the course that Hubel and Wiesel taught us postdocs and students in the early 1970s. However, I should hasten to add that feature detection as an explicit goal of visual processing was never discussed. Hubel and Wiesel appeared to assume that understanding the receptive field properties of visual neurons would eventually explain perception, and that further discussion would be superfluous.

In light of all this, it will seem odd that the rest of the book is predicated on the belief that these widely accepted ideas about how the visual brain works are wrong. The further conclusion that understanding what we see based on learning more about the responses of visual neurons is likely to be a dead end might seem even stranger. Several things conspired to sow seeds of doubt after years of enthusiastic, if remote, acceptance of the overall program that Hubel and Wiesel had been pursuing. The first flaw was the increasing difficulty that they and their many acolytes were having when trying to make sense of the electrophysiological and anatomical information that had accumulated by the 1990s. In the early stages of their work, the results obtained seemed to beautifully confirm the intuition that vision entails sequential and essentially hierarchical analyses of retinal image features leading to the neural correlates of perception (see Figure 7.3). The general idea was that the luminance values, spectral distributions (colors), angles, line lengths, depth, motion, and other features were abstracted by visual processing in the retina, thalamus, and primary visual cortex, and subsequently recombined in increasingly complex ways by neurons at progressively higher stages in the visual cortex. These combined representations in the extrastriate regions of the visual system would lead to the perception of objects and their qualities by virtue of further activity elicited in the association cortices in the occipital lobes and adjacent areas in the temporal and parietal lobes.

A particularly impressive aspect of Hubel and Wiesel's observations in the 1960s and 1970s was that the receptive field properties of the neurons in the lateral geniculate nucleus of the thalamus could nicely explain the properties of the neurons they contacted in the input layer of the primary visual cortex, and that the properties of these neurons could explain the responses of the neurons they contacted at the next higher level of processing in V1. The neurons in this cortical hierarchy were referred to as "simple," "complex," and "hypercomplex" cells, underscoring the idea that the features abstracted from the retinal image were progressively being put back together in the cortex for the purpose of perception. Although I doubt Hubel and Wiesel ever used the phrase, the rationale for the initial abstraction was generally assumed to be engineering or coding efficiency.

These findings also fit well with their anatomical evidence that V1 is divided into iterated modules defined by particular response properties, such as selectivity for orientation (see Figure 7.4) or for information related to the left or right eye (see Figure 6.2A). By the late 1970s, Hubel and Wiesel had put these several findings together in what they called the "ice cube" model of visual cortical processing

(Figure 7.7). The suggestion was that each small piece of cortex, which they called a "hyercolumn," contained a complete set of feature-processing elements. But as the years passed and more evidence accumulated about visual neuronal types, their connectivity, and the organization of the visual system, the concept of a processing hierarchy in general and the ice cube model in particular seemed as if a square peg was being pounded into a round hole.

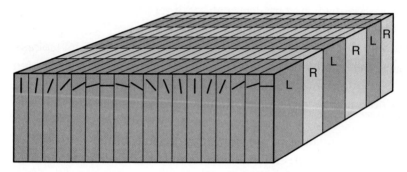

Figure 7.7   The ice cube model of primary visual cortical organization. This diagram illustrates the idea that units roughly a square millimeter or two in size (the primary visual cortex in each hemisphere of a rhesus monkey brain is about 1,000 square millimeters) each comprise superimposed feature-processing elements, illustrated here by orientation selectivity over the full range of possible angles (the little lines) and comapping with right and left eye processing stripes (indicated by L and R; see Figure 6.2A). (After Hubbel, 1988)

A second reason for suspecting that more data about the receptive field properties of visual neurons and their anatomical organization might not explain perception was the mountain of puzzling observations about what people actually see, coupled with philosophical concerns about vision that had been around for centuries. Taking such things seriously was a path that a self-respecting neuroscientist followed at some peril. But vision has always demanded that perceptual and philosophical issues be considered, and the cracks that had begun to appear in the standard model of how the visual brain was supposed to work encouraged a reconsideration of some basic concerns. One widely discussed issue was the question of "grandmother cells," a term coined by Jerry Lettvin, an imaginative and controversial neuroscientist at MIT who liked the role of intellectual and (during the Vietnam War era) social provocateur. If the features of retinal images were being progressively put back together in neurons with increasingly more complex properties at higher levels of the brain, didn't this imply the existence of nerve cells that would ultimately be ludicrously selective (meaning neurons that would respond to only the retinal image of your grandmother, for example)? Although the question was facetious, many people correctly saw it as serious. The ensuing debate was further stimulated by the discovery in the early 1980s of neurons in the association areas of the monkey brain that did, in fact, respond specifically to faces (an area in the human temporal lobe that responds selectively to faces has since been well documented). A related question concerned the binding problem. Even if visual neurons don't generate perceptions by specifically responding to grandmothers or other particular objects (which most people agreed made little sense), how are the various features of any object brought together in a coherent, instantaneously generated perception of, for example, a ball that is round, chartreuse, and coming at you in a particular direction from a certain distance at a certain speed (think tennis). Although purported answers to the binding problem were (and still are) taken with a grain of salt, most neuroscientists recognized that such questions would eventually need to be answered. Although a lot of my colleagues were not very interested in debates of this sort, I had always had a weakness for them and was glad to see these issues raised as serious concerns in neuroscience. After all, I had been a philosophy major in college and had left clinical medicine because I wanted to understand how the brain worked, not just how to understand its maladies or the properties of its constituent cells.

By the mid-1990s, I began to be bothered by another philosophical issue relevant to perception that was ultimately decisive in reaching the conclusion that mining the details of visual neuronal properties would never lead to an understanding of perception or its underlying mechanics. Western philosophy had long debated about how the "real world" of physical objects can be "known" by using our senses. Positions on this issue had varied greatly, the philosophical tension in recent centuries being between thinkers such as Francis Bacon and René Descartes, who supposed that absolute knowledge of the real world is possible (an issue of some scientific consequence in modern physics and cosmology), and others such as David Hume and Immanuel Kant, who argued that the real world is inevitably remote from us and can be appreciated only indirectly. The philosopher who made these points most cogently with respect to vision was George Berkeley, an Irish nobleman, bishop, tutor at Trinity College in Dublin, and card-carrying member of the British "Empiricist School." In 1709, Berkeley had written a short treatise entitled *An Essay Toward a New Theory of Vision* in which he pointed out that a two-dimensional image projected onto the receptive surface of the eye could never specify the three-dimensional source of that image in the world (Figure 7.8). This fact and the difficulty it raises for understanding the perception of any image feature is referred to as the inverse optics problem.

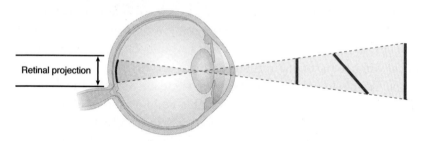

Figure 7.8   The inverse optics problem. George Berkeley pointed out in the eighteenth century that the same projected image could be generated by objects of different sizes, at different distances from the observer, and in different physical orientations. As a result, the actual source of any three-dimensional object is inevitably uncertain. Note that the problem is not simply that retinal images are ambiguous; the deeper issue is that the real world is directly unknowable by means of any logical operation on a projected image. (After Purves and Lotto, 2003)

In the context of biology and evolution, the significance of the inverse problem is clear: If the information on the retina precludes direct knowledge of the real world, how is it that what we see enables us to respond so successfully to real-world objects on the basis of vision? Helmholtz was aware of the problem and argued that vision had to depend on learning from experience in addition to the information supplied by neural connections in the brain determined by inheritance. However, he thought that analyzing image features was generally good enough and that a boost from empirical experience (empirical experience, for him, was what we learn about objects in life through trial-and-error interactions) would contend with the inverse problem. This learned information would allow us to make what Helmholtz referred to as "unconscious inferences" about what an ambiguous image might represent. Some vision scientists seemed to take Helmholtz's approach to the inverse optics problem as sufficient, but many simply ignored it. The problem was rarely, if ever, mentioned in the discussions of vision I had been party to over the years. In particular, I had never heard Hubel and Wiesel mention it or saw it referred to in their papers.

At the same time, I was increasingly aware in the 1990s, as anyone who delves into perception must be, of an enormous number of visual illusions. An illusion refers to a perception that fails to match a physical measurement made by using an instrument of some sort: a ruler, a protractor, a photometer, or some more complex device that makes direct measurements of object properties, therefore evading the inverse problem. In using the term *illusion* the presumption in psychology texts and other literature is that we usually see the world "correctly," but sometimes a natural or contrived stimulus fools us so that our perception and the measured reality underlying the stimulus fail to align. But if what Berkeley had said was right, analysis of a retinal image could not tell the brain anything definite about what objects and conditions in the world had actually generated an image. It seemed more likely that *all* perceptions were equally illusory constructions produced by the brain to achieve biological success in the face of the inverse problem. If this was the case, then the evolution of visual systems must have been primarily concerned with solving this fundamental challenge. Surprisingly, no one seemed to be paying much attention to this very large spanner that Berkeley had tossed into logical and analytical concepts of how vision works.

I didn't have the slightest idea of how the visual wiring described by Hubel and Wiesel and their followers might be contending with the inverse problem. But I was pretty sure that it must be by means of a very different strategy from the one that had been explicitly or implicitly dominating my thinking (and most everyone else's) since the 1960s. If understanding brain function was going to be possible, exploring how vision contends with the inverse problem seemed a very good place to start.

8

Visual perception

The minor epiphany in 1997 that started me thinking in earnest about a possible answer to the inverse problem was a picture. I was listening to a lecture by David Somers, a visiting postdoc from MIT. I don't remember the subject, but in the course of the talk, he showed a popular stimulus that his mentor, psychologist Ted Adelson, had created. The picture was an abstract pattern of light and dark areas, similar to others that had been around in some form since the late nineteenth century, showing that regions with the same amount of light coming from them could produce quite different perceptions of brightness. Adelson had published the stimulus in a paper that proposed a model seeking to account for the effect in terms of the receptive field properties of "midlevel" visual cortical neurons. I had read the paper, but when looking at this pattern again, it dawned on me that its abstract nature was obscuring the fact that the elements of the pattern that looked lighter were, in effect, being presented as if they were in shadow, and the physically identical elements that looked darker were presented as if they were well illuminated. The upshot was the thought that the brightness differences that the physically identical patches elicited corresponded to the empirical meaning of images that human beings had experienced from time immemorial.

What struck me as important was not that the information in a scene influenced perception of some particular part of it— Helmholtz, the Gestalt psychologists, and pretty much everyone else had recognized that—but that an accumulation of this information in the visual brain arising from trial-and-error behavior in response to retinal stimuli provided a general way of getting around the inverse problem. Looking at perception in this way also suggested why the standard ideas about what visual neurons were doing had not been able to explain what we see, and perhaps pointed the way to understanding what the connectivity of visual neurons was actually accomplishing. I was excited enough about the idea that perceptions might be determined in this way to go back to the lab after the lecture and seek out Mark Williams, a postdoc who was especially skilled in computer graphics (a methodology that was far more challenging then than it is with today's user-friendly software). I sketched a crude scene and tried to explain what I thought might be going on. Although what I said couldn't have made much sense, it was enough to get him interested.

Williams had sought me out in despair a few months earlier, soon after he had completed his Ph.D. in Pat Goldman's lab at Yale. He had become dispirited by the prolonged effort and the modest scientific payoff of what he had been doing, and was thinking of leaving neuroscience. Given his talent in graphics and deep knowledge of human neuroanatomy, I urged him to figure out a career path in which he could use these skills, and offered him a position to give it a try. He agreed and began collaborating with Len White, putting together an interactive digital atlas of the human brain (being a poorly qualified teacher in this area, I very much supported the project). Following this conversation, Williams made a few digital pictures that suggested promising ways of exploring this take on perception. Within a few weeks, he had created a series of computer programs that we could use to test the idea that the brightness values people see are determined by linking patterns of luminance to perceptions that would have facilitated successful behavior in response to the scene in question.

Brightness seemed a very good place to start. Of the basic qualities that define visual perception (brightness, lightness, color, geometric form, depth, and motion), the simplest and, arguably, most important are brightness and lightness. Given the niche a species occupies, an animal might have little or no color vision, see forms poorly, have little or no ability to discriminate depth by stereopsis, or even be effectively blind (for example, some species of bats, moles, and mole rats). But all animals with vision must discriminate light and dark. *Lightness* and *brightness* are the names given to these subjective qualities, and, as with all perceived visual qualities, they arise from the neural processing of some physical aspect of light energy. *Lightness* is the term applied to the appearance of a surface that reflects light, such as a piece of paper, whereas *brightness* refers to the appearance of a source that emits light, such as a light bulb. (Although *brightness* is sometimes loosely used to refer to both qualities, this distinction between *brightness* and *lightness* is important, as described in Chapter 10.) The physical parameter associated with both lightness and brightness is luminance, which refers to the overall amount of light energy measured by a photometer. However, lightness, brightness, or any other visual quality can't be measured directly, which creates a problem for psychophysicists. These subjective qualities can be evaluated only by asking an observer to report the appearance of one object relative to the appearance of another, or in terms of some imagined scale that covers the range of that quality (for example, a scale of 1–100 on which subjects rate the dimmest [1] to brightest [100] stimulus).

Because it would presumably behoove us to see the world as it "really is," a logical expectation is that lightness or brightness should closely follow the luminance of a stimulus—that is, the more the amount of light falling on some region of the retina, the lighter or brighter the corresponding region in the scene should appear. But this expectation is not met, and we never see the world as a photometer measures it. Figure 8.1 illustrates a simple example of the disconnect between lightness or brightness and luminance. The two central circles have the same measured luminance and, therefore, return the same amount of light to the retina. Nevertheless, they are perceived differently. The patch on the darker background appears lighter or brighter than the same patch on the lighter background. In the jargon of the field, this phenomenon is called simultaneous lightness/brightness contrast.

For anyone who has doubts about this demonstration (or thinks it might be fudged), it is easy enough to create the effect by cutting and pasting a few pieces of appropriately light and dark papers cut from a magazine. And if you think that the apparent difference in the appearance of the targets in Figure 8.1 is not really big enough to worry about, Figure 8.2 shows that such effects can be made much stronger by presenting patches with the same luminance in a more natural (information-rich) scene. Generating dramatically different perceptions of surfaces that are physically the same in terms of the amount of light they return to the eye is not new. In 1929, German psychologist Adhemar Gelb created a room in which he hung a piece

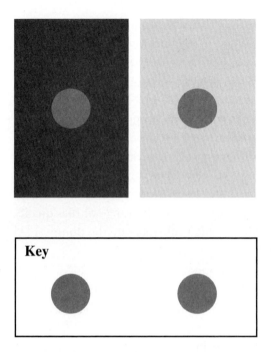

Figure 8.1   A simple example of the discrepancies between the measured luminance of surfaces and the perception of lightness or brightness. When measured by a photometer, the circles in the dark and light surrounds give identical readings. However, the target circle on the dark background looks lighter or brighter than the target on the light background. The key shows that when the targets are presented on the same background, they appear the same. This similarity of appearance does not mean that we are seeing "physical reality" in the key and an "illusion" in the top part of the figure, as explained later in the chapter. (After Purves and Lotto, 2003)

of paper illuminated by a hidden lamp. Depending on the level of the illumination of the objects in the rest of the room, the piece of paper could be made to look nearly white or nearly black, despite the same amount of light reaching the observer's eye from the paper.

During the next year, Williams and I, with the help of Alli McCoy, a bright and energetic premed student, put together enough evidence to write a couple of papers on how subjects equalized the brightness or lightness of test patches in scenes with different empirical meanings (Figure 8.3). The question was whether the adjustments people made reflected these different meanings, and invariably they did. The implication was that accumulated experience viewing the world, not the luminance relationships in the retinal image or a direct

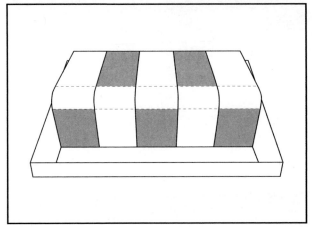

Figure 8.2 A more dramatic example of how surfaces that return the same amount of light to the eye can look very different. As shown in the key below, the indicated patches all have exactly the same luminance; however, in a scene they can look either much darker or much lighter than expected from their measured properties. Again, don't be misled by the idea that because the patches in the neutral context look the same, the key is somehow revealing "reality;" in both instances, the brain is making up the shades of gray seen, and the shades in the key are no more "real" than those in the scene above. (After Purves and Lotto, 2003)

analysis of these physical features, determined the different lightness or brightness seen in response to the same patch in a light versus a dark surround in Figure 8.1—or in any other stimulus of this general sort, such as Figure 8.2.

(A)

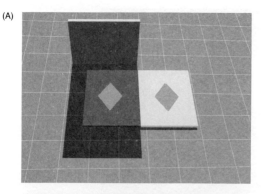

(B)

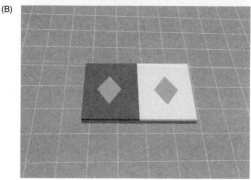

(C)

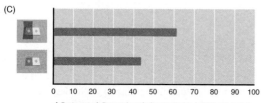

0   10   20   30   40   50   60   70   80   90   100

Adjustment of diamond on dark surround to brightness match

Figure 8.3   Example of scenes used to explore the idea that lightness and brightness are determined by accumulated empirical experience. A) Test targets (the diamonds) of the same luminance located in surrounds similar to those in Figure 8.1 but presented with information indicating that the left diamond lies in shadow and the right diamond in light. B) The same test targets and surrounds presented as if they were intrinsic (painted-on) components of surfaces under the same illumination. C) Difference in the lightness of the identical test targets and surrounds reported by observers. The upper bar shows the average adjustments made to equate the lightness of the diamond on the dark surround with that of the diamond on the light surround in response to scene (A). The lower bar shows the adjustment made to equalize the appearance of the two diamonds in response to scene (B). (After Williams, et al., 1998. Copyright 1998 National Academy of Sciences, U.S.A.)

We assumed that contending with inverse problem was the underlying cause of these effects. The challenge presented to biological visual systems is easy enough to appreciate if one considers how the luminance values on the retina are generated by physical objects and real world conditions. Three fundamental aspects of the physical world determine the amount of light that reaches the eye: the illumination of objects, the amount of the light reflected from object surfaces, and the amount of the reflected light subtracted by the atmosphere (or any other material, such as water or glass) in the space between objects and the observer. As shown in Figure 8.4, the relative contributions of these factors are always entangled in the light stimuli; an infinite number of combinations of illumination, reflectance, and intervening transmittance can create the same value of retinal luminance. As a result, the visual system cannot logically analyze the retinal image to decipher how these factors have actually been combined to generate a particular luminance. Because successful behavior requires responses that accord with the physical sources of a stimulus, this inherent uncertainty presents a daunting obstacle to the evolution of vision. If the lightness/brightness values that we see were simply proportional to the luminance values, the result would be a useless guide to behavior. However, if our sense of lightness and brightness is generated empirically—by trial-and-error accumulation of information that reflected successful behavior in response to the patterns of light arising from natural scenes—this manifestation of the inverse problem could be circumvented (as Figure 8.4 indicates, it can't be "solved" in a logical sense). Given that the visual system can't determine the nature of objects and conditions in the world by logical operations on image features, the most likely— and perhaps only—alternative would be to do so empirically.

The word *empirical* has several related meanings, but the applicable definition in this context is "evidence obtained by trial and error." If retinal images can't specify their sources, one way around the inverse problem would be to link images to their generative sources according to behavioral success by tallying up what a given image had turned out to be in past experience. We (or other visual animals) could then use this information to produce perceptions of the world that would work in behavioral terms, even though what we would see as a result would never correspond with physical measurements. This scheme sounds

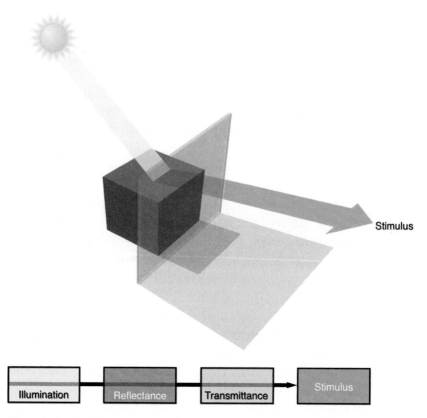

Figure 8.4   The inevitable entanglement of the physical factors (the illumination, the reflectance of surfaces, and the transmittance of the atmosphere)
that create luminance values in visual stimuli. As a result, the physical sources
of a luminance value on the retina cannot be known by any logical operation
carried out on the retinal image. (From Lotto and Purves, 2003)

crazy, but gradually storing up this empirical information in the evolving circuitry of the visual system over millions of years would effectively resolve the inverse problem. A corollary is that the way we
actually perceive the world would provide a powerful way of testing
whether this is indeed the strategy that the human brain is using in
vision and the other sensory modalities.

Happily, what we see in various circumstances has been well documented over the years, providing plenty of grist for the mill of this or
any proposed explanation of vision. The documentation falls into two
general categories: all the visual illusions whose bases have been
debated during the last century or more, and all the less well-known

(some would say boring) psychophysical functions that vision scientists have amassed. The focus in what follows is on the more intriguing "gee whiz" category of so-called illusions, setting aside psychophysical functions for the time being. Work on these phenomena had, for the most part, been relegated to psychology; most neuroscientists (myself included until the late 1990s) regarded illusions as a by-product of the neuronal interactions that neurophysiologists were unraveling, and of no great interest in their own right. There were some exceptions to this parochial mindset. A few outstanding vision scientists—Horace Barlow at Cambridge is perhaps the preeminent example—combined work on visual physiology (Barlow described receptive fields in the frog retina about the same time Kuffler described them in the cat retina) with studies of perception. But this was not the tradition that I had been exposed to. A case in point is Hubel and Wiesel's magisterial retrospective published in 2006 called *Brain and Visual Perception,* a book that combines their key papers with commentaries on what they did and why they did it. Despite the title, the book has no index entry for perception, no definition of this term in the glossary or elsewhere, and little or no discussion in the papers or commentaries of the parallel work on perceptual issues that was going on in psychology and psychophysics during the 25 years of their collaboration.

Perception is defined as the conscious product of sensory processing, given to us as the qualities that we attribute back to objects and conditions in the world (the way things look, sound, feel, smell, and taste). This concept, however, is a very limited description of what is really occurring in the nervous system. Although we think in terms of vision, audition, bodily sensations, taste, and smell, we are oblivious to the output of many other sensory modalities; the information produced by sensory systems that monitor muscle length and tension that Hunt and Kuffler had studied, blood pressure, blood gas levels, and a host of other parameters critical to survival never enters our awareness.

Earlier in the twentieth century, the Gestalt school of psychology was the most popular approach to explaining perception. The most famous representatives of this way of thinking were a trio of German psychologists who emigrated to the United States in the 1920s and 1930s: Max Wertheimer, Wolfgang Köhler, and Kurt Kofka. The assumption of the Gestaltists was that many perceptual phenomena arise because we tend to see scenes in terms of organized forms or wholes. (*Gestalt* means "form" or "shape" in German.) The organizational rules that presumably applied to a complete scene compared to its elements would be different and would therefore lead to different appearances. This way of thinking had the merit of taking the complete content of scenes into account, but the Gestalt laws that emerged seemed little more than another way of describing the perceptual effects. Although Gestalt psychologists documented many challenging phenomena (the Gelb room is a good example), they didn't link their "laws" to biology. As a result, enthusiasm for this approach even among psychologists had begun to fade by midcentury.

By the 1950s, explaining the discrepancies between lightness or brightness perceptions and luminance was done primarily on the basis of electrophysiology. Based on the evidence that had begun to emerge from such studies, one way of rationalizing the effects elicited by stimuli such as the standard simultaneous brightness contrast stimulus in Figure 8.1 was to consider them incidental consequences of visual processing—a price paid, so to speak, for achieving some larger goal of the visual system. For example, the central region of the retinal output cells' receptive fields was known to have a surround with the opposite functional polarity, an organization that presumably enhances the detection of the edges (Figure 8.5). As a result, retinal neurons whose receptive fields lie over a light–dark boundary but with their central regions either within or without the dark region of an image would be expected to generate action potentials at a somewhat different rate. This physiological observation made it attractive to suppose that the patch on the dark background in Figure 8.1 looks lighter or brighter than the patch on the light background because of this difference in the retinal output to the rest of the visual brain. Kuffler, who had reported the organization of receptive fields of neurons in the cat retina in 1953, avoided this sort of assertion, probably because he felt that tying receptive field properties directly to perception was a dangerous game, which it is. An appreciation of Kuffler's conservative philosophy might be one of the reasons why his protégés Hubel and Wiesel paid little attention to visual perception.

However, the perceptions elicited by other stimulus patterns show that simultaneous brightness contrast effects such as those in Figure 8.1 are not an incidental consequence of retinal processing. In

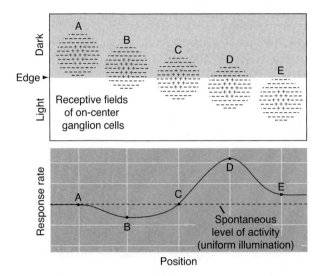

Figure 8.5   The receptive field properties of retinal ganglion cells and their output to higher visual processing centers. The upper panel shows the receptive fields of several retinal neurons (ganglion cells) overlying a light–dark boundary. The lower panel shows the different firing rates of ganglion cells as a function of their position with respect to the boundary. In principle, this difference might cause the lightness/brightness contrast effects in Figure 8.1 as an incidental consequence of neural processing at the input stages of vision. (After Purves and Lotto, 2003)

Figure 8.6, the four target patches on the left are surrounded by relatively more higher-luminance (lighter) territory than lower-luminance (darker) territory; nevertheless, they appear lighter than the three targets on the right, which are surrounded by relatively more lower-luminance (darker) territory. Although the average luminance values of the surrounding contexts of the patches in the stimulus are effectively opposite those in standard simultaneous brightness stimulus in Figure 8.1, the lightness/brightness differences elicited are about the same in both direction and magnitude. The effect of the scene in Figure 8.2 would also be difficult to explain as an incidental consequence of retinal neuron receptive field properties.

The difficulty of explaining lightness or brightness perceptions is compounded by the absence of any cortical region dedicated to processing luminance. And although a relatively close link exists between light intensity and the rate of action potential generation by neurons in the retina and visual thalamus, this relationship breaks down when

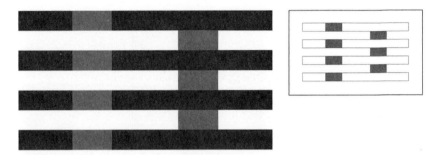

Figure 8.6   A stimulus pattern that elicits perceptual effects that, as described in the text, undermine explanations of lightness/brightness contrast based on the properties of neurons at the input level of the visual system. The key indicates the physically identical target patches of interest. (From Purves and Lotto, 2003)

neurons are tested in the "higher" processing stations of the visual system. Neurons in the visual cortex generally respond only weakly to changes in light intensity. Yet these neurons represent the more complex aspects of stimuli, as evidenced by the selective receptive field properties of complex and hypercomplex neurons, and even neurons that respond to particular objects (see Chapter 7). Although neuronal activity at these higher cortical levels must be generating what we see, the representation of luminance—the most basic of the visual qualities—seems to have been lost in the shuffle.

How, then, could an empirical perspective ever hope to make sense of the perceptually perplexing and physiologically messy relationship between luminance and lightness or brightness? Because explanations of perceptual qualities in wholly empirical terms are a recurrent theme in the chapters that follow, it may help to define how, in principle, this strategy could account for the strange way that we perceive brightness and lightness. The standard stimulus in Figure 8.7A (and Figure 8.1) is, on empirical grounds, consistent with the two identical targets being physically similar surfaces under the same amount of light (Figure 8.7B), or the two physically different surfaces under different amounts of light (Figure 8.7C). In a wholly empirical framework the frequency with which these possibilities occur in human experience would have influenced the evolution of the relevant visual circuitry accordingly, creating neural connectivity in the visual brains of successive generations that reflected the consequences of trial-and-error responses to natural stimuli. Some members of each generation, through inheritance and random genetic variation, would possess visual circuitry that generated perceptions that facilitated behavioral responses to the sources of stimuli a little more effectively than the circuitry in other members of the cohort. These individuals would, on average, be slightly more successful in life and reproduce a little more effectively. As a result, neural connections that linked the relevant visual stimuli to operationally useful perceptions and behavior would gradually wax in the visual brains of the population. Perceptions arising in this way would correspond not to any particular feature of the stimulus, but to the lightness or brightness perceptions that had proved the best link to successful behavior in the past. By contending in this way with the inverse problem, identical targets in different surrounds in Figure 8.7A would look differently light or bright because the perceptions of the sources needed to generate successful behavior would necessarily be different (compare Figures 8.7B and 8.7C).

To make the biological rationale for this way of generating perceptions more concrete, think of a collection of surfaces—such as pieces of paper—having a range of physical compositions, with some reflecting more light and some less light when measured under the same illumination. If vision's goal is to distinguish objects that are physically the same or different in any circumstance, it would be of little or no use to perceive the luminance values generated by the surfaces of the papers. For example, imagine cutting one of the papers in half and placing the two halves at random somewhere in the room you are sitting in; the identical halves would likely end up in places where they would be illuminated differently and would therefore return two different luminance values to the eye. Conversely, the luminance of two physically different papers could be the same under different conditions of illumination, again making their luminance useless as a means of distinguishing their actual characteristics. Given that it is biologically useful to see the two papers as members of the same class (imagine them to be edible, and thus something of biological value), human ancestors who made this distinction a little better because of their slightly different visual connectivity would reproduce a little more successfully as a result of this advantage. Eventually, the visual system of humans or other species would discriminate surfaces in different contexts quite well on this empirical basis,

(A) Standard simultaneous brightness contrast stimulus

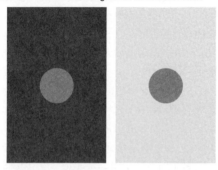

(B) Similar surfaces under similar illuminates

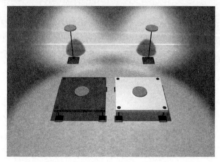

(C) Different surfaces under different illuminates

Figure 8.7   Understanding the relationship between luminance and lightness/brightness in empirical terms. A) A standard simultaneous lightness/brightness contrast stimulus, such as the one in Figure 8.1. (B) and (C) show the two different physical situations that would both have produced the stimulus in (A). See text for further explanation. (From Purves and Lotto, 2003)

despite the inherently uncertain meaning of luminance values in retinal images.

Assuming that this conception of vision made sense, the next step for us was to test its validity in a serious way. At least two possibilities came to mind. Because many stimuli elicit peculiar lightness or brightness effects, we could test whether each of these particular puzzles has a plausible explanation in empirical terms. Alternatively, we could test this interpretation of vision more directly by asking whether human experience with luminance values in retinal images in relation to objects and conditions in the world accurately predicts the lightness or brightness people end up seeing in response to any given stimulus. Pursuing these goals during the next few years depended critically on Beau Lotto, a postdoc who arrived in the lab in 1998. I had been lucky the day Jeff Lichtman appeared in my office in 1974, and I was equally lucky when Beau showed up. Like Tim Andrews, Lotto had done his doctorate in developmental neurobiology in the United Kingdom, and had come with the idea of pursuing some developmental issue in vision. But by the time he arrived, my interest in development had faded and the effort directed at visual perception was growing. After some initial discussion about what to do, he threw himself into the work on perception, and he had all the skills needed to push things along in new and imaginative ways. Similar to Lichtman, Lotto grasped the nub of conceptual or technical problems right away, and he had the intelligence and tenacity to solve them. He was (and is) as much an artist as a scientist, and his ability to make visual tests and demonstrations was as good as Williams's, who was about to leave to start a company (Williams eventually became a successful entrepreneur in graphics and iPod and iPhone applications, many of them related to medical education and the brain).

Because we had no idea how to acquire or analyze data that could serve as a proxy for human visual experience with luminance or anything else, we started with the easier task of showing that some of the most perplexing brightness effects had plausible empirical bases. The first phenomenon we studied was known as Mach bands (Figure 8.8). Ernst Mach was a nineteenth-century German physicist whose name has been immortalized in the unit given to the speed of sound (Mach 1), and whose philosophical opposition to the reality of atoms helped set the stage for Einstein's paper on Brownian motion in 1905. Much like his contemporaries Helmholtz and Maxwell, Mach was deeply

interested in vision, and in 1865 he described a band of increased
darkness at the onset of a luminance gradient and a band of increased
lightness at its offset that had no physical basis (*gradient* here means
a gradual transition).

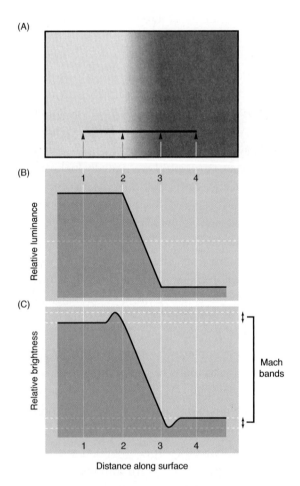

Figure 8.8   Mach bands. The stimulus that elicits this effect is shown in the
upper panel (A); the middle (B) and lower (C) panels show, respectively, the
physical (photometric) and perceptual profiles of the stimulus. A luminance gradient between a uniformly lighter and darker region causes people to see a
band of maximum lightness at position 2 and a band of maximum darkness at
position 3. The luminance values in the middle panel are those on the printed
page in (A) and show that these bands have no basis in physical reality. (After
Purves and Lotto, 2003)

Another example that involves gradients is the Cornsweet edge effect (Figure 8.9). Tom Cornsweet, an accomplished psychologist, instrument designer, and psychophysicist, described the effect that bears his name about a century after Mach's report. In this case, a gradient from gray to black, when opposed to a gradient from gray to white (forming the "edge" in the name of the effect), causes adjacent gray territories identical in luminance to look different. The surface adjoining the gradient going from gray to black looks darker than the surface that abuts the gradient going from gray to white.

Given their physical and mathematical bent, both Mach and Cornsweet proposed that these odd effects are incidental consequences of physiological interactions in the retina, and elaborated detailed models (and complicated equations) to indicate how retinal processing could lead to the perceptions. Despite a lot of further work by psychologists and psychophysicists, these theories remained in the literature as the best explanations. Lotto and I argued that both Mach bands and the Cornsweet edge effect were more likely to be based on the link between the luminance values in the stimuli that Mach and Cornsweet devised and the real-world sources of these stimuli that observers would always have experienced. In the case of Mach bands, we showed that the curved surfaces that humans routinely see produce light gradients with *actual* regions of increased and decreased luminance at their offset and onset; these are the familiar highlights and lowlights produced by the physical effects of curvature on reflected light (Figure 8.10A). In the same vein, we showed that when the elements of a Cornsweet edge stimulus are present in natural scenes, the adjacent territories *do* return physically different luminance values to the eye (Figure 8.10B).

The empirical interpretation of these perceptual phenomena is that human experience interacting with the natural sources of the stimuli that Mach and Cornsweet described would have influenced the evolution of visual circuitry according to the success or failure of the behavioral responses to these stimuli. As a result, the lightness and brightness values generated by visual system circuitry would, over evolutionary time, gradually become determined by the relative success of that perception in dealing with the relevant sources. In consequence the lightness/brightness values seen would reflect this

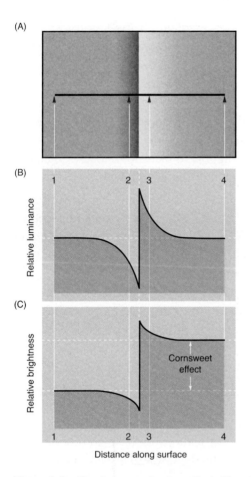

Figure 8.9   The Cornsweet edge effect. The stimulus is shown in the upper panel (A); the middle (B) and lower (C) panels show the physical and perceptual profiles of the stimulus, as in Figure 8.8. Despite the identical luminance values of the territories adjoining the two gradients (2 and 3), the entire territory between 1 and 2 looks darker than the territory between 4 and 3. (After Purves, et al., 1999)

operational success and not luminance values per se, explaining the effects that Mach and Cornsweet had described.

Although Lotto and I were pleased with what we took to be clever empirical explanations of phenomena that had puzzled people for a long time, the general response of vision scientists was that these accounts could not be taken seriously because they were not linked to the enormous body of physiological information about visual neurons that everyone assumed would soon explain perception. This complaint

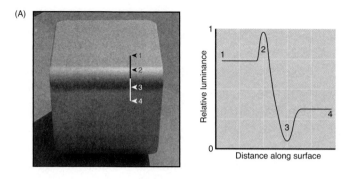

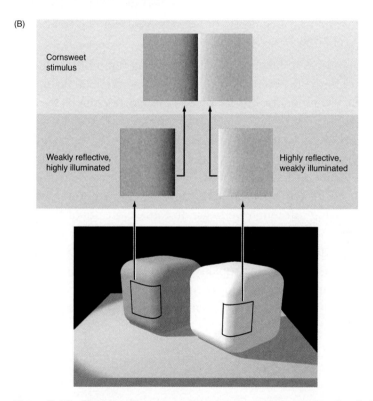

Figure 8.10   Real-world sources of the Mach and Cornsweet stimuli. A) Photograph of an aluminum cube in sunlight. Physical highlights and lowlights adorn the curved surfaces of real-world objects, as indicated in the accompanying photometric measurement. Note the similarity of the luminance profile here to the perceptual profile of Mach bands in Figure 8.8. B) As shown in this diagram of two cubes with curved surfaces made of different material and under different amounts of illumination, the components of the Cornsweet stimulus also arise routinely in natural scenes. When they do, the flat surfaces made of different material that abut the curvatures that generate the luminance gradients have different luminance values. (From Purves and Lotto, 2003)

was frustrating: If we were right, the relationship between the properties of visual neurons and perception could never be explained in the logical framework that everyone had assumed would give the answers to such puzzles. What we see is certainly caused by light stimuli that activate neurons in the primary and higher-order regions of visual cortex. But if a wholly empirical strategy is the brain's operating principle, then the visual system was a welter of neuronal connections built up historically over evolutionary (and individual) time according to all the factors that go into the determination of successful behavior. Not surprisingly, people whose careers were founded on a logically understandable hierarchy of sensory processing did not welcome this different idea of how brains work.

In an empirical conception of vision, making sense of the neurophysiology underlying perception would ultimately depend on understanding the evolutionary history of the human nervous system in terms of behavior. And that goal seemed impossible, at least in the short term. If we had a proxy for visual experience with some limited aspect of the natural world—such as human experience with luminance—we could test whether that body of empirical information predicted some of the lightness and brightness perceptions that we had been studying. Although we began to think more about this possibility, it was not clear how to proceed in 1999. The easier path was to add more evidence to our claims by examining the perceptions elicited by some other visual quality in empirical terms, asking if other longstanding puzzles could be reasonably accounted for in this way. Color seemed to be the visual quality best suited to this purpose.

9

Perceiving color

Color is a fascinating perceptual quality, and almost everyone interested in vision during the last few centuries—Newton in the seventeenth and eighteenth centuries; Young, Maxell, Helmholtz, Mach, and even Goethe in the nineteenth century; and a host of investigators in the twentieth century—has wrestled with it. The extraordinary effort to rationalize color vision is ironic because color vision is not very important biologically: Many animals (dogs and cats, for example) have little color vision, and people with most forms of "color blindness" have only minor problems getting along in life.

What is color vision, and why have some species (including ours) gone to the trouble of evolving it? The perceptions of lightness and brightness discussed in the last chapter concern what we see in response to the overall amount of light that reaches the eye; by definition, these qualities are the perceptual responses elicited by this physical aspect of light stimuli. Seeing in shades of gray that range from black to white occurs when the energy in stimuli is more or less evenly distributed across the light spectrum (Figure 9.1A); seeing in color is the perceptual quality generated in us, and presumably many other visual animals, when the energy in a light stimulus is unevenly distributed, as Isaac Newton first showed in the late seventeenth century (Figure 9.1B). However, Newton recognized that neither light nor objects are colored. Our brains generate the colors we see for reasons of biological advantage, just as brains make up the qualities of all our other perceptions. If you have doubts about this assertion, consider the perception of pain. The sensation we perceive when we accidentally touch a hot stove is not a feature of the world but a sensory quality that leads to useful behavior. The entirely subjective sense of pain makes us remove the hand before much damage is done and teaches us to be cautious of such objects in the future.

A

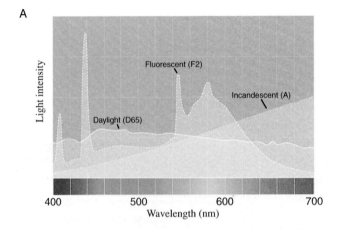

B

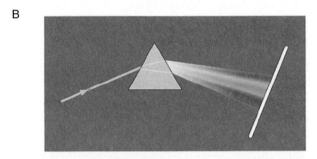

Figure 9.1 White light spectra and their decomposition. A) The spectra of daylight, fluorescent light, and light from an ordinary light bulb (a tungsten filament) do not elicit the perception of color when reflected from surfaces that preserve their relative uniformity. Although the energy in these several sources varies across the spectrum, it is distributed evenly enough so that all three human cone types are stimulated to about the same extent. B) As Newton showed, "white light" can be broken up by a prism into component spectra. The components elicit perceptions of color because they stimulate the three cone types in the human retina to different degrees. (From Purves and Lotto, 2003)

    The advantage of color vision is more subtle and less important than distinguishing objects on the basis of luminance, which explains the relatively innocuous consequences of color-deficient vision. Nevertheless, the advantage of seeing color is not hard to appreciate. Although any visual animal can distinguish boundaries that depend on relative amounts of light, animals with little or no color vision are unable to distinguish surface boundaries that do not differ in luminance but only in the distribution of spectral power. A visual animal that can identify boundaries based on both these qualities of light by seeing colors in addition to light and dark will be better able to distinguish and classify objects in the environment, such as predators or prey. Animals that have not evolved color vision presumably did not need this extra visual aid as much as the animals that did. As might be expected, animals that don't see color well are typically nocturnal, or exist in an ecological niche that does not offer much need for this real but relatively modest visual advantage.

By the end of the nineteenth century, Young, Helmholtz, and Maxwell had argued convincingly that the spectral differences that Newton first described generate color perceptions by differentially activating three different receptor types. In the twentieth century, the postulated receptors were found to be three cone types distinguished by a different light-absorbing pigment in each. Helmholtz's idea that these three receptor types determine the colors we see is called the *trichromatic theory* of color vision. Sensible as his idea seems, the trichromatic theory is only half right: The three human cone types certainly play a critical role in color vision (witness the deficient color vision that occurs when one or more of the cone pigments is altered or missing), but they are only part of story.

The idea that stimulation of the three cone types is a sufficient explanation of color vision ran into trouble almost right away. Ewald Hering, Helmholtz's contemporary and sometime nemesis, pointed out that some aspects of the colors we see are not readily understood in terms of retinal receptor types. Hering's main objection was that humans with normal color vision perceive red to be an opponent color to green, and blue to be an opponent color to yellow. Although observers can see or imagine a transition from red to yellow through a series of intermediates without entertaining any other primary color perception, no parallel perception—or conception—exists for getting from red to green without going through blue or yellow; and there is no way of getting from blue to yellow without going through red or green. Moreover, humans perceive a particular red, green, blue, and yellow to be unique, meaning that we see one particular value in each of these four color categories as having no admixture of any other colors (see Figure 10.1). This unique quality is different than the way we see other colors. There is, for example, no unique orange or purple, orange always being seen as a mixture of red and yellow, and purple as a mixture of blue and red. Because the different activation of the three retinal cone types by light offers no explanation of these perceptual phenomena, trichromacy theory is, Hering argued, an incomplete account of how color sensations are generated. Even today there is no consensus about why we see these four "primary" color categories.

The central question that has intrigued everyone from neurobiologists to philosophers is why we see the colors in the way we do. Coming up with a plausible answer has been difficult, primarily because color vision entails so many puzzles. In addition to Hering's concern about the perception of opponent colors, it has long been known that color perceptions don't match the physical measurements made by a spectrophotometer. Similar to the lightness and brightness effects described in Chapter 8, when two target patches return the same spectrum to the eye but are surrounded by regions that reflect different distributions of light energy, the color perceptions elicited by the two targets are changed (Figure 9.2). This phenomenon is called *simultaneous color contrast*. Just as puzzling are effects that seem the opposite of color contrast: the context surrounding two patches that reflect different spectra to the eye can elicit similar color perceptions, a phenomenon called *color constancy* (Figure 9.3). Thus, a banana looks yellow and an apple looks red whether the fruits are observed in the "bluish" light from the sky, the "reddish" light of sunset, or the "yellowish" light that comes directly from the sun. In each circumstance, the distribution of light energy reaching the eye from the surfaces of the same objects is quite different. These puzzling effects underscored Hering's argument that color vision cannot be explained simply by the activation of retinal receptors. It was these anomalies in particular that got Beau Lotto and me thinking about how color perception might be explained in empirical terms.

Several first-rate psychologists and psychophysicists had studied color contrast and constancy in detail in the early decades of the twentieth century. But the person whose name had become most closely associated with attempts to rationalize these anomalies was Edwin Land. Land was a Harvard College dropout whose genius had already been established by his many contributions to photography (the invention of the "instant camera" among them). By the late 1950s,

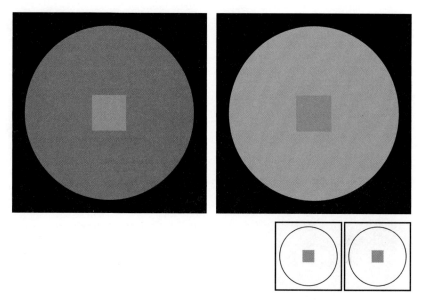

Figure 9.2   A typical color contrast stimulus. The two central squares are returning identical spectra to the eye and are perceived as the same color when presented on the same background, as shown in the key. However, the target in a reddish surround looks yellowish, and it looks reddish in a yellowish surround. (Again, don't be misled into thinking that the key is revealing the "real colors" of the patches; every color perception is a product of the brain, not something that exists in reality.) (After Purves and Lotto, 2003)

he had amassed a substantial fortune from the success of the Polaroid Corporation that he had founded in 1937. Intrigued by some of the mysteries of color vision he had to contend with in color photography, Land took on the challenge of color contrast and constancy and in so doing stimulated a revival of interest in these phenomena. He did so primarily through a series of demonstrations that he presented with considerable fanfare (Figure 9.3). The best known of these used a collage of physically different papers illuminated by three independently controlled light sources that provided long-, middle-, and short-wavelength light, respectively. (Because the collages resemble the work of Dutch artist Piet Mondrian, such stimuli are often referred to as "Land Mondrians.") Land first adjusted each of the three light sources to some value and then determined the spectral return from one of the surfaces in the array of papers (such as a surface that looked yellowish to observers when it was illuminated by all three lights). Under exactly the same illumination, he showed that another

(A)

(B)

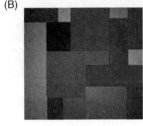

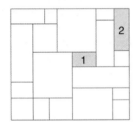

Figure 9.3   Land's demonstration of color constancy. A) The appearance of a collage when illuminated by a particular mixture of long-, medium-, and short-wavelength light. Because patches 1 and 2 (see the key) have different physical properties, they return different spectral stimuli to the eye and, as expected, look differently colored. B) The same collage as it would have appeared after readjusting the three sources so that the spectrum returned from patch 2 is now the same spectrum initially returned from patch 1. Despite the drastic change in the spectral returns from the patches, patch 2 continues to look reddish and patch 1 continues to look yellowish. (After Purves and Lotto, 2003)

patch in the collage (such as one that looked reddish) provided a substantially different spectrum reaching the eye, as would be expected from the different physical properties of the two papers. Land then readjusted the three illuminators so that the "yellowish" paper now provided exactly the same spectral return to observers as the spectrum that had originally come from the "reddish" paper. Common sense suggests that the "yellowish" paper in the collage should now have looked like the "reddish" paper in the previous condition of illumination. However, the yellowish patch of paper continued to look more or less yellow, and the reddish patch (which was also returning a different spectrum to the eye under the new conditions) continued to look more or less red.

Although these kinds of results had been demonstrated repeatedly since the nineteenth century, Land's fame and the provocative way he stated his case created a vigorous debate. "We have come to the conclusion," Land opined in a manner that was bound to irritate the vision scientists who had been working on these issues, "that the classic laws of color mixing conceal great basic laws of color vision" (Land, 1959). Much of the ensuing controversy focused on whether Land had shown anything new. He had not, but his demonstrations emphasized that color vision was deeply complicated by the fact that the entire scene is somehow relevant to the perceived color of any part of it (as is true for stimuli that elicit perceptions of lightness or brightness) and that understanding color vision would require understanding these phenomena.

Land tried to explain the interaction of the spectral information in a scene by postulating that contrast and constancy are based on neural computations that entail the ratios of cone activity multiplied by the illuminant at each point in the retinal image. Other investigators more familiar with the physiology of neuronal interactions suggested that the basis of these contextual effects was probably input-level adaptation to the predominant spectral contrasts in a scene. But other stimuli elicit color perceptions that are inconsistent with the predictions of Land's theory or theories based on adaptation. For example, in Figure 9.4, the spectra of the targets (the circles in [A] and the crosses in [B]) are identical in all four panels, and the spectra of the orange and yellowish backgrounds in the upper and lower panels are also the same. However, the targets on the left in both the upper and lower panels appear to be about the same color, as do the targets on the right. These similar color perceptions of the targets occur despite the fact that the surrounds are opposite in the upper and lower panels. This confound is a similar to the puzzle presented by the grayscale stimulus in Figure 8.6.

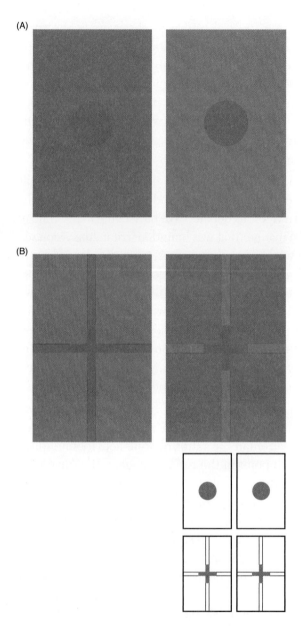

Figure 9.4   A further challenge to rationalizing color contrast and constancy. A) A typical color contrast stimulus in which two identical central targets embedded in different spectral surrounds appear differently colored (see Figure 9.2). B) In this configuration, the central targets are physically the same as those in the upper panels (see the key), but the backgrounds have been switched. Despite this switch, the apparent colors of the left and right targets are about the same in the upper and lower panels. (From Purves and Lotto, 2003)

This bit of history provides some sense of the challenges facing anyone who wants to say something sensible about why we see the colors we do. But based on the inverse problem and the accounts that we had been able to give for otherwise puzzling brightness effects such as Mach bands and the Cornsweet edge effect (see Chapter 8), Lotto and I felt pretty sure that color perceptions must also be explainable in empirical terms. After all, the entanglement of the factors generating luminance values illustrated in Figure 8.4 applies equally to spectral information: The conflation of the illuminant, the surface properties of objects, and the subtraction of light by the atmosphere would make the real-world sources of color spectra just as uncertain as the sources of the overall amount of light in visual stimuli. Because no logical operation on the distribution of light energy in retinal images—the basis of color vision—could specify what real-world objects and conditions had given rise to the relevant stimulus, color perceptions would also have to be determined by accumulated trial and error experience.

For a number of reasons—including the contentious nature of the subject and its inherent complexity—developing an empirical theory of color vision is considerably more challenging than rationalizing the relationship between luminance and lightness/brightness in these terms. Nevertheless we began exploring the idea that the odd way we see color is the result of trial-and-error experience with spectral relationships accumulated over evolutionary time. Our supposition was that color contrast and constancy, similar to the lightness/brightness phenomena described in Chapter 8, must arise from linking spectral stimuli and physical sources according to behaviors that worked in the past. If so, then the nature of this accumulated empirical information should predict color perceptions.

We started out testing this idea by having subjects adjust the perceived color of a target on a neutral background until it matched an identical target on a color background, measuring in this way the perceptual change induced by the color of the surround with stimuli similar to the example in Figure 9.2. The relationships that we determined in this way did not rule out Land's theory, adaptation, or other schemes based on interactions among neurons sensitive to color stimuli. But they did show that color contrast could be equally well understood as the outcome of a visual strategy in which color perceptions are generated according to past experience. Much of this work depended on Lotto's extraordinary skill as an artist, and Figure 9.5 shows him with one of the public installations that he produced to demonstrate the ideas described here.

Figure 9.5   Beau Lotto in 2008 with one of his public art installations in London. (Courtesy of Beau Lotto)

Figure 9.6 shows how an empirical explanation of color perception would work in the case of a color contrast stimulus. The genesis of the light coming from the targets in the standard color contrast stimulus in Figure 9.6A is not knowable directly; as indicated in the cartoons in Figures 9.6B and 9.6C, many different combinations of reflectance, illumination, and transmittance could have produced the same stimulus. Accordingly, the biological strategy outlined in Chapter 8 for determining what lightness or brightness an observer should see in response to a given luminance (see Figure 8.7) is equally applicable to color: Contend with the inverse problem by color perceptions based on neural circuitry determined by operationally successful behavior in

A

B

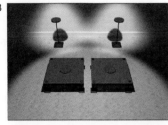

C

Figure 9.6 An empirical explanation of color contrast. The standard color contrast stimulus in (A) could have been generated by physically identical targets on physically different backgrounds under white light illumination (as in [B]), or physically different surfaces on physically similar backgrounds under illuminants with different color spectra (as in [C]). In an empirical framework, the different color appearance of the identical targets in (A) is a result of operational links made over evolutionary and individual time between the inevitably uncertain meaning of the spectral characteristics of the retinal image and experience with the relative success of behavioral responses made to the same or similar stimuli in the past. (After Purves, et al., 2001)

the past. The circuitry needed to facilitate successful behavior would have been shaped according to on how often the stimulus in Figure 9.6A was generated by the situation in Figure 9.6B versus 9.6C, or any of the many other combinations of surface properties, illumination, and other factors that could produce the stimulus in Figure 9.6A. The different color appearance of the two targets would be the perceptual signature of this way of contending with the inverse problem.

Beau and I gained more confidence in this explanation of color perception by manipulating spectral patterns that enhanced or diminished color contrast by making the characteristics of a scene either more or less consistent with different possible bases for the light coming from target patches (Figure 9.7). For example, when identical targets (the central squares) are presented on backgrounds that comprise a variety of tiles with spectra designed so that the two arrays are likely to be under "red" and "blue" illumination, respectively, the apparent color difference between the targets is relatively marked (Figure 9.7A). Conversely, when Lotto made the contextual information consistent with the arrays under the same illumination, the apparent color difference of the physically identical targets decreased (Figure 9.7B). Because the average spectral content in the left and right scenes is the same, these effects cannot be explained by neuronal interactions at the input stages of the visual system or adaptation. However, in empirical terms, the color perceptions elicited by the same targets in different contexts make good sense. The waxing of circuitry underlying successful behaviors over time would have incorporated the empirical fact that when the light spectrum coming from two patches is the same but the illumination is different, the objects are always physically different surfaces. However, when the light from the patches is the same and they are under the same illumination, experience would have taught that the objects are physically the same. The rationale for the differences we actually see (color contrast effects) is that two things that *are* different must *look* different for behavior to be successful, even when they are returning exactly the same light spectrum to the eye.

The same argument can explain color constancy in empirical terms. Most people had assumed (and many still do) that color constancy is an explicit goal of vision, meaning that bananas, apples, or any other object should continue to look their respective colors in all circumstances so that we can more readily identify the objects (the usual example given is identifying the ripeness of fruits, an ability that would presumably have evolutionary value). We thought it was more likely that color constancy is just another indicator that the colors we see are generated empirically, with color contrast and constancy being two sides of the same coin. If the information in a scene is consistent with accumulated experience of interactions with two targets that return

A

Scene consistent with different chromatic illuminants

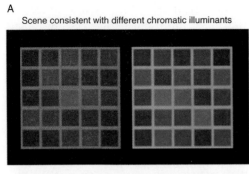

B

Scene inconsistent with different chromatic illuminants

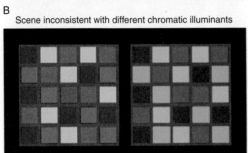

Figure 9.7 Altering color perception by manipulation of empirical significance. A) A stimulus in which a physically identical central tile is presented in the context of other tiles whose spectral returns are all consistent with "reddish" illumination of the left panel and "bluish" illumination of the right panel. B) The same scene as (A), but with the central target surrounded by tiles whose spectral returns are consistent with similar illumination of both left and right panels. See text for explanation. (After Lotto and Purves, 2000. Copyright 2000 National Academy of Sciences, U.S.A.)

the same spectra to the eye but turn out to be different physical surfaces, then they will appear differently colored (color contrast); conversely, if the information in a scene is inconsistent with this possibility, then they will tend look similarly colored, even if the two targets return different spectra to the eye (color constancy).

Figure 9.8 sums up this interpretation of color contrast and constancy. Although the spectra coming from the central squares on the two faces of the cube in Figure 9.8A are identical, their colors are different because the information in the stimulus is what we humans would have experienced when interacting with differently reflective surfaces under different illuminants. Conversely, the same two targets on the faces of the cube in Figure 9.8B appear relatively similar even though the spectra coming from the targets are different, because this is the information we would always have experienced when interacting with physically similar surfaces under different illumination. In both instances, seeing color in this way would have fomented successful behavior.

(A)   (B)

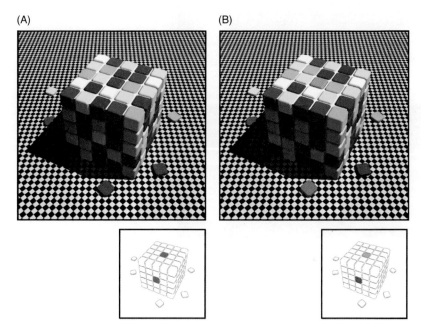

Figure 9.8   Color contrast and constancy as consequences of the same empirical strategy of color vision. The two panels demonstrate the effects on the color appearance of surfaces when two physically similar target surfaces (as in [A]) or two physically different target surfaces (as in [B]) are presented in the same context. Because the spectral information in the scene is consistent with different intensities of light falling on the upper and lateral surfaces of the cube, a color contrast effect is seen in (A) and a color constancy effect is seen in (B). The appearances of the relevant target surfaces in the same neutral context are shown in the keys. (From Purves and Lotto, 2003)

A further twist in trying to make this argument compelling was to create scenes designed to appear as they would if seen under differently "colored" lights (Figure 9.9). Lotto thought correctly that this further trick would generate even more dramatic effects on color appearances. As indicated by the keys, the "blue" tiles on the top of the cube in Figure 9.9A are physically identical to the "yellow" tiles on the cube in Figure 9.9B (both sets of tiles appear gray when the varied spectral information in the surrounding scene is masked). Therefore, spectrally identical patches are being made to look either blue or yellow. Because blue and yellow are color opposites, the gray patches are being made to appear about as far apart in color perception as possible. This is a powerful demonstration of empirically determined color contrast. Conversely, the "red" tiles on the cube in Figure 9.9A appear similar to the "red" tiles on the cube in Figure 9.9B, even though the spectra coming from them are very different, as indicated in the keys. This latter comparison is a powerful demonstration of color constancy. Our explanation was that the pertinent empirical information had been built into the visual system circuitry as a consequence of species and individual experience in the world, and that the resulting contrast and constancy effects were behaviorally useful responses to the relevant stimuli.

The empirical framework we had cooked up also had the potential to explain another longstanding puzzle in color vision: why mixing different amounts of short-, medium-, and long-wavelength light at a given level of intensity generates most but not all color perceptions. Because such mixing can produce any distribution of power across the light spectrum, in principle, it should produce any color perception. But seeing colors such as browns, navies, and olives requires adjusting the context as well. These color perceptions occur only when the intensity of the light coming from the relevant region is low compared to the light coming from the other surfaces in a scene. For example, brown is the color seen when the luminance of a surface that otherwise looks orange is reduced compared to the luminance of the surfaces around it. The inability to produce brown, navy, and olive color perceptions by light mixing without changing the overall pattern of light intensities in a scene explains why the spectrum coming from the tile on the top of cube in Figure 9.8 looks brown, whereas the same light spectrum coming from the face in shadow looks orange.

Taken together, all this evidence seemed to indicate convincingly that the colors we see are not the result of the spectra in retinal images per se, but result from linking retinal spectra to real-world objects and conditions discovered empirically through the consequences of behavior. Regarding color contrast and constancy effects as "illusions" was not only wrong, but also missed the biological significance of these phenomena and the explanatory power of understanding color vision in empirical terms.

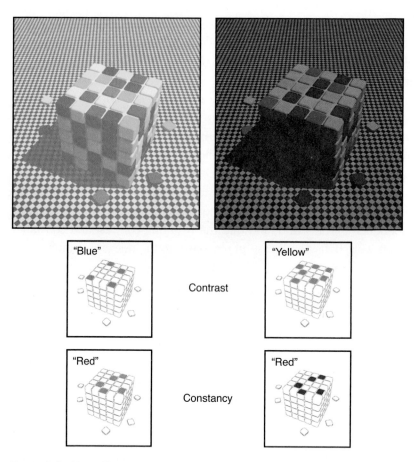

Figure 9.9   The effects on color perception when the same or physically different targets are presented in scenes consistent with illumination by differently
"colored" light sources (the quotes are a reminder that because colors are a
result of visual processing, neither light nor the tiles can properly be called colored). Except for the targets, all the spectral information has been made consistent with "yellowish" illumination in the scene on the left, and with "bluish"
illumination in the scene on the right. See text for further explanation. (After
Purves and Lotto, 2003)

Although Lotto and I were sold on the idea, the response from the
vision scientists who we thought might be converted was a collective
yawn, or worse. Anonymous reviewers invariably gave the papers we
wrote a hard time, granting agencies were unenthused, and even local
colleagues showed only polite curiosity about what we were doing.
This generally negative reaction came in two flavors. Visual physiologists were unimpressed because we said nothing about how any of this

could be related to the properties of visual neurons and implied that what they were doing was off the mark. Indeed, if visual circuitry had been selected empirically because it linked retinal images to behaviors that worked in response to a world that was directly unknowable by means of light, it seemed doubtful that the logical framework that people had long relied on would ever explain vision.

An example of the opposition was the reaction to a talk I gave on this work on lightness/brightness and color to a small group in southern California—the Helmholtz Club—in 2001. Francis Crick, who had turned to neuroscience in the late 1970s after revolutionizing the understanding of genetic inheritance, had started the club after becoming especially interested in vision as a means of understanding consciousness. Although he was not a physiologist, Crick was certainly a consummate reductionist. When I argued to the 20 or so club members in attendance that vision would need to be understood in this way, Crick became apoplectic, exclaiming heatedly that he wanted to know about the nuts and bolts of visual mechanics, not some vague theory based on the inverse problem and the role of experience. My protest that there was a lot of evidence supporting the interpretation based on what we actually see—and that the discovery of genes and ultimately DNA had depended on vague ideas about inheritance in the late nineteenth century—didn't change his summary dismissal of the whole idea.

The second sort of opposition came from psychologists who were equally unenthusiastic, but for different reasons. Psychologists and psychophysicists had worked on brightness and color for decades, advancing a variety of theories, often mathematically based, that purported to explain at least some of these phenomena. Many of the critiques we suffered pointed out that we were not sufficiently familiar with this abundant literature and that we had failed to rebut (or sometimes even mention) these other explanations. Although this complaint was partly true, the main objection seemed to be that in coming at these issues from a different tradition, we lacked the necessary credentials to intrude in this arena and failed to appreciate its history, complexities, and conventional wisdom. The most this group seemed willing to grant was that we (meaning Lotto) had made some very pretty pictures.

Whatever the reasons for antipathy, it was apparent that explaining perceptual phenomena empirically without data about accumulated human experience that could be used to predict specific perceptions would not carry the day. But it wasn't clear how to get or analyze such information. Moreover, in going this route we would have to confront the way the brain organizes the lessons of experience in the perceptual space of a visual quality, meaning the way values of lightness, brightness, color, or some other perceptual quality are subjectively related. Having already irritated the physiologists and psychologists, a further effort in this more abstract direction—which could be considered toying with the structure of "mind"—would invite the philosophers to pile on as well.

10

The organization of perceptual qualities

Despite the subjective nature of perception, there are several ways to assess how the brain organizes a given quality. In general, understanding the organization of a perceptual category depends on determining how the quality behaves when a relevant stimulus parameter is systematically varied. For example, perceptual qualities such as lightness, brightness, or color can be evaluated in terms of the least amount of light energy that can be seen as the wavelength of a stimulus is varied, as the background intensity is altered, as a pattern is changed, or as some other aspect of the stimulus is permutated. Another approach is to determine the least noticeable differences between stimuli that can be perceived, for example between two stimuli that differ slightly in luminance. Yet another way of determining the organization of perception is simply to measure the time it takes people to react to a systematically varied stimulus. Assessing the organization of perceptual qualities in vision or some other sensory modality in these ways has established many "psychophysical functions."

These efforts to make the organization of what we perceive scientifically meaningful date from about 1860, when German physicist and philosopher Gustav Fechner decided to pursue the connection between what he referred to as the "physical and psychological worlds" (thus the modern term "psychophysics"). In vision, the goal of psychophysics is to understand how the brain organizes the perception of qualities such as lightness, brightness, color, form, and motion. For lack of a better phrase, this organization is often referred to as the *perceptual space* pertinent to the quality in question. The significance of such studies for us was clear enough. We would have to rationalize the configuration of the perceptual spaces indicated by the psychophysical functions that had been painstakingly determined over the years in empirical terms if this is indeed the strategy that evolved to contend with the unknowability of the world through the senses.

Although Lotto and I recognized the need to move beyond the ad hoc explanations of the particular perceptual phenomena discussed in Chapters 8 and 9, we found plenty of reasons to hesitate in pursuing this goal. Dozens of attempts to describe the organization of the perceptual space of lightness, brightness, or color and had already been made without much success but plenty of debate. Part of the difficulty in reaching a consensus about the organization of any perceptual space is the complexity of the problem. As the diagram in Figure 10.1 implies, although it was convenient to separate the discussion of color and lightness/brightness in the preceding chapters, the lightness/brightness values we see are intertwined with color perceptions, and vice versa. The spectral distribution of light affects the perceptions elicited by luminance, and luminance affects the colors elicited by light spectra. The entanglement of these and other visual qualities had been amply documented in the classic colorimetry experiments described later in the chapter, and this presents a major obstacle to rationalizing the perceptual space in Figure 10.1 in empirical or any other terms. As a result, many vision scientists took a dim view of plumbing these perceptual details as a useful way forward.

Nonetheless, it seemed worth a try. We began by looking more specifically at the organization of the full range of lightness/brightness perceptions in the absence of any influence from hue or color saturation (the central axis of perceptual space shown in Figure 10.1). Many studies of lightness and brightness had shown that the organization of these perceptual qualities in relation to the physical intensity of light (luminance) is far more complex than the vertical straight line in the diagram implies, as should already be apparent from the anomalies described in Chapters 8 and 9. The most notable student of this issue was Harvard psychophysicist Stanley Stevens, who worked on these topics from about 1950 to 1975. Stevens wondered how a simple white light stimulus made progressively more intense over a range of physical values is related to the lightness/brightness values that subjects reported. To determine a person's sense of lightness or brightness in response to a particular stimulus in a quantitative way, he asked subjects to rate the intensity they experienced in response to

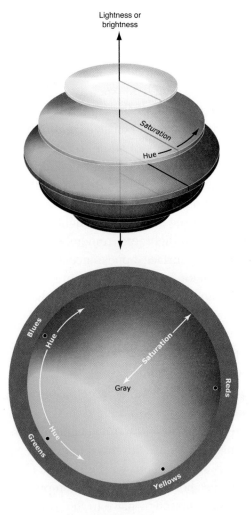

Figure 10.1   Diagram of the human perceptual space for lightness/brightness and color. The vertical axis in the upper panel indicates that any particular level of light intensity evokes a particular sense of lightness or brightness; movements around the perimeter of the relevant plane correspond to changes in hue (changes in the apparent contribution of red, green, blue, or yellow to perception), and movements along the radial axis correspond to changes in saturation (changes in the approximation of the color to the perception of a color-neutral gray). The lower panel is looking down on one of the planes in the central region of this space. The four primary color categories (red, green, blue, and yellow) are each characterized by a unique hue (indicated by dots) that has no apparent admixture of the other three categories of color perception (a color experience that cannot be seen or imagined as a mixture of other colors). (After Lotto and Purves, 2003)

a given stimulus on an imaginary scale on which 0 represented the least intense perception in response to the stimuli in the test series and 100 represented the most intense (Figure 10.2). Stevens, and subsequently other investigators using different methods, found that when the test stimuli were sources of light (such as a light bulb or the equivalent), doubling the luminance of the stimulus did not simply double the perceived brightness; the change in brightness was greater than expected at lower values of stimulus intensity, but less than expected for higher values of the test light (see the red curve in Figure 10.2). However, Stevens obtained a different result when the same sort of test was carried out using a series of papers that ranged from light to dark instead of using an adjustable light source. In this case, the reported perceptions of lightness (remember that *lightness* is the term used to describe surface appearances and *brightness* describes the appearance of light sources) varied more or less linearly with the measured luminance values generated by the reflective properties of the papers (see the green line in Figure 10.2). If vision operates according to an empirical strategy, we should be able to explain these different psychophysical functions (often called "scaling functions") for light sources and surfaces.

Lotto and I, together with Shuro Nundy, a graduate student who eventually wrote his thesis on these issues, started thinking about how these observations could be explained as more general manifestations of the empirical arguments that we had advanced to explain particular effects such as Mach bands and the Cornsweet edge (see Chapter 8). Because we had very little idea how to do this, our attempts were bumbling and seemed—no doubt correctly—naive to the psychophysicists in this field. Nevertheless, we were reasonably sure that we could explain Stevens's observations in terms of human experience linking luminance values with the underlying sources discovered according to the results of behavior. Because the intensities of sources of light are generally greater that the intensities of reflected light (surfaces typically absorb some of the light that falls on them), the discovery of these facts through behavioral interactions would have shaped the organization of the visual circuitry underlying the perceptual spaces of lightness and brightness differently. The incorporation of this empirical information in visual system circuitry would explain the difference in the psychophysical functions for sources and surfaces illustrated in Figure 10.2.

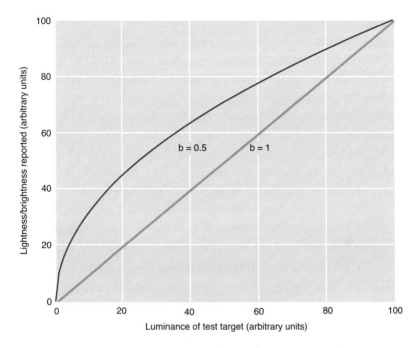

Figure 10.2   A summary of the psychophysical (or scaling) functions observed in studies carried out by Stevens and others in which test targets are presented as either light sources or surfaces over a range of intensities (luminance values). When the target stimuli are sources of light (the sort of stimulus that elicits a sense of brightness), the subjective rankings that subjects report tend to track along the red curve with an exponent (b) of about 0.5. However, if the stimuli are a series of surfaces (such as pieces of paper) that reflect different amounts of light, then the subjective rankings of lightness track closer to the green line with an exponent that approaches 1. (After Nundy and Purves, 2002. Copyright 2002 National Academy of Sciences, U.S.A.)

Although this framework seemed simplistic, it provided a way of exploring the relative lightness and brightness values people would expect to see if the organization of perceptual space for these qualities is indeed determined empirically. The initial efforts to test this interpretation seemed promising. Shuro presented subjects with a series of test patches ranging from black to white on a computer screen set against a featureless gray background, and compared the responses in this circumstance to those elicited by the same patches presented in a scene as objects lying on an illuminated tabletop. The observers' responses shifted from nonlinear in response to patches presented on the featureless background (consistent with the patches being light sources) to more nearly linear when the information in the tabletop scene implied that the patches were surfaces reflecting light from an illuminant.

These and other observations Shuro made supported the conclusion that the psychophysical functions for lightness and brightness that Stevens and others had established (see Figure 10.2) could be explained in the same empirical terms that we had used to rationalize the phenomena discussed in Chapter 8. The argument in this case was that the same amount of light coming from a test patch would, through behavioral interactions, have turned out to be sometimes a source of light and sometimes light reflected by a surface. For simple physical reasons (the absorption of some light by reflecting surfaces), light sources typically return more light to the eye than is reflected from surfaces in the rest of any scene. The consequences of these facts for successful behavior would have shaped the perceptual spaces of lightness and brightness differently, leading to the psychophysical functions illustrated in Figure 10.2.

Another perplexing set of psychophysical observations that needed to be explained was the results obtained in classical colorimetry experiments. Colorimetry involves careful measurements of the way the physical characteristics of spectral stimuli are related to the colors that observers see, carried out in the simplest possible testing conditions (Figure 10.3). Because color perception includes the subsidiary qualities of hue, saturation, and color brightness (see Figure 10.1), colorimetry testing shows how these qualities interact. For example, colorimetry indicates how changes in hue affect the brightness and how changes in brightness affect the perception of hue. Such studies can be thought of as a more complete way of examining the organization of color space diagrammed in Figure 10.1. As indicated in Figure 10.3, the stimuli in such experiments are typically generated by three independent light sources producing long, middle, and short wavelengths projected onto half of a frosted-glass screen set in a black matte surround. A test light, usually having a single wavelength, is projected onto the other half of the diffuser. The subject is then asked to adjust the three light sources until the color of the two halves of the disk appears the same. Alternatively, the experimenter might gradually vary the wavelength or intensity of the test stimulus and ask subjects to report when they first noticed a difference between the appearances of the two halves (a test of "just noticeable differences").

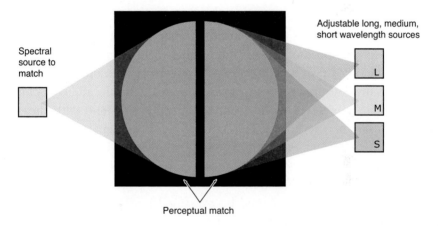

Figure 10.3   Colorimetry testing. See text for explanation. (After Purves and Lotto, 2003)

Psychophysical functions determined by colorimetry testing can be generated by color matching tests or color discrimination tests. In color matching tests, such studies showed that saturation varies in a particular way as a function of luminance (the so-called Hunt effect); that hue varies as a function of stimulus changes that affect saturation (the Abney effect); that hue varies as a function of luminance (the Bezold–Brücke effect); and that brightness varies as a function of stimulus changes that affect both hue and saturation (the Helmholtz–Kohlrausch effect). In color discrimination tests, the ability to distinguish "just noticeable" differences in hue, saturation, or color brightness also varies in a complex way as a function of the wavelength of the test stimulus.

Explaining these interdependent functions is fiendishly complicated. The few brave souls who tried—psychophysicists such as Leo Hurvich, Dorothea Jameson, Walter Stiles, and Gunter Wyszecki— had developed models based on well-documented knowledge of how the three human cone types respond to light of different wavelengths, coupled with assumptions about the neuronal interactions that do (or might) occur early in the visual pathway. Although such models were extraordinarily clever, they were ad hoc explanations and accounted for one or another colorimetric phenomenon without providing any biological rationale. The supposition was that these perceptual phenomena were simply incidental consequences of light absorption by the different receptor types and subsequent visual processing.

The complexity of these colorimetric phenomena (and the resulting literature) presented a daunting challenge to anyone trying to understand color vision. Reticent as we were to take up this arcane subject, if we wanted to show that an empirical interpretation of color space had merit, we would eventually need to contend with these observations in colorimetry. The framework for this departure was the same as that we had used to explain the lightness/brightness functions that Stevens and others had determined (see Figure 10.2). Even under the simple conditions of colorimetry, testing the perception of hue, saturation, and brightness should be explained by the typical relationships among the characteristics of light stimuli that humans would have incorporated in visual circuitry through feedback from trial-and-error behavior. As in lightness and brightness scaling, the colorimetry functions that had been described would be signatures of the empirical strategy that evolved to contend with the inverse problem. Therefore, if we knew the approximate nature of human experience with the physical attributes underlying hue, saturation, and color brightness, we should be able to predict these functions and greatly strengthen the empirical case.

Although we could infer the nature of human experience with object surfaces and light sources that would have caused some of the anomalies apparent in color contrast and constancy (see Chapters 8 and 9), or even the lightness and brightness scaling function in Figure 10.2, intuition was useless when it came to colorimetry functions. The only way to determine the experience that we thought might underlie these functions was to examine spectral relationships in a large database of natural scenes. The way forward in this project was the result of much hard work by two postdoctoral fellows who had recently come to the lab from mainland China, Fuhui Long and Zhiyong Yang. Like many other American scientists working during the last decade or two, I owe an enormous debt of gratitude to the Chinese educational system, which was, and is, turning out wonderful students as the quality of education at all levels in the United States continues to decline. The skills in mathematics, computer programming, and image analysis that Long and Yang brought with them were simply not part of the intellectual equipment of most students trained in this country.

Long was a quiet young woman whose English left much to be desired when she started out in 2001. But her retiring demeanor concealed a vivid intelligence and a determination that I had rarely encountered in homegrown students, and she was never shy in the daily arguments that play a big part in any new project. Long had received a Ph.D. in computer science and electrical engineering, and had been a postdoc in the Department of Electronic and Information Engineering at Hong Kong Polytechnic University, where she had worked on image processing and computer vision. As a result, she was facile with a wide range of technical approaches to problems in image analysis, and her purpose in coming to my lab was to gain some knowledge about the biological side of things. After some false starts, we settled on analyzing color images as a means of understanding the human color experience underlying colorimetry functions. Long eventually collected more than a thousand high-quality digital photographs of representative natural scenes for this purpose (Figure 10.4A) and wrote the computer programs needed to extract the physical characteristics associated with hue, saturation, and color brightness at each point in millions of smaller image patches taken from these pictures (Figure 10.4B). Our assumption was that this database would fairly represent the spectral relationships that humans had always experienced, and we hoped this information would enable us to predict the classical colorimetry functions.

An example of this approach is explaining the results of color discrimination tests in empirical terms. Psychophysicists Gunter Wyszecki and Walter Stiles generated the function shown in Figure

A

B

Figure 10.4 Examples of the scenes used to assess human color experience (A). By analyzing millions of small patches from hundreds of these natural scenes (B), we could approximate the accumulated experience with the spectral variations and relationships that we thought must underlie the psychophysical functions determined in colorimetry testing. (After Long, et al., 2006. Copyright 2006 National Academy of Sciences, U.S.A.)

10.5A in the late 1950s; it indicates the human sensitivity to a perceived color change as the wavelength of a stimulus is gradually varied. The resulting function is anything but simple, going up and down repeatedly over the range of wavelengths that humans perceive. Where the slope of the function is not changing much, people make finer color discriminations: Relatively little change in the wavelength of the test stimulus is needed before the subject reports an apparent difference in color. However, where the slope is steep a relatively large change in wavelength is needed before observers see a color difference. The reason for these complex variations in sensitivity to wavelength change was anybody's guess.

(A)

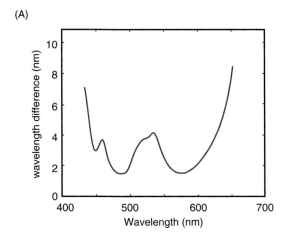

(B)

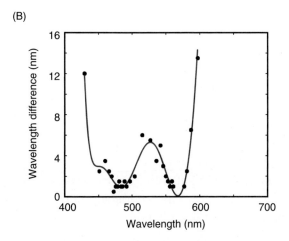

Figure 10.5 Prediction of a colorimetric function from empirical data. A) Graph showing the amount of change in the wavelength of a stimulus needed to elicit a just noticeable difference in color when tested across the full range of light wavelengths. B) The function predicted by an empirical analysis of the spectral characteristics in a database of natural scenes (see Figure 10.4). The black dots are the predictions from the empirical data, and the red line is a best fit to the dots. (The data in [A] were drawn from Wyszecki and Stiles, 1958; [B] is from Long, et al., 2006, copyright 2006 National Academy of Sciences, U.S.A.)

Predicting this function on empirical grounds depends on the fact that the light experienced at each point in stimuli arising from natural scenes will always have varied in particular ways determined by the qualities of object surfaces, illumination from the sun, and other characteristics of the world. For instance, middle-wavelength light is more likely than short- or long-wavelength light at any point because of the spectrum of sunlight and the typical properties of object surfaces. The wavelengths arising from a scene will also have varied as the illumination changes because of the time of day, the weather, or shadowing from other objects. These and myriad other influences cause the spectral qualities of each point in visual stimuli to have a typical frequency of occurrence (see Figure 10.4). Humans will always have experienced these relationships and, based on this empirical information, will have evolved and refined visual circuitry to promote successful behavior in response to the inevitably uncertain sources of light spectra. If the organization of the perceptual space for color has been determined this way, then the effects of this routinely experienced spectral information should be evident in the ability of subjects to discriminate color differences as the wavelength of a stimulus is varied in colorimetry testing. Figure 10.5 compares this psychophysical function with the function predicted by analyzing the spectral characteristics of millions of patches in hundreds of natural scenes. Although certainly not perfect, the psychophysically determined function in Figure 10.5A is in reasonable agreement with the function predicted by the empirical data in Figure 10.5B. The other colorimetry effects mentioned earlier could also be predicted in this same general way.

Just as important as rationalizing scaling and colorimetry functions was to show how the empirical molding of the perceptual spaces for lightness and brightness could explain the contrast and constancy phenomena considered in Chapters 8 and 9, and Zhiyong Yang focused on this task. Yang had received a doctorate in computer vision from the Chinese Academy of Sciences. Before coming to the lab, he had done postdoctoral work with David Mumford at Brown working on pattern theory and with Richard Zemel at the University of Arizona on probabilistic models of vision. Like Long, his remarkable skills were well suited to extracting empirical information from scenes and figuring out how we could use this information in a more quantitative way to explain otherwise puzzling perceptions. Although

Lotto and I had suggested empirical explanations for lightness/brightness contrast stimuli, they had not been taken very seriously, and the perceptual effects generated by many other patterns of light and dark remained to be accounted for. A good example is the stimulus in Figure 8.4 (White's effect), a pattern that produces a perception of the lightness or brightness of targets that is opposite the effect elicited by the standard brightness contrast stimulus (see Figure 8.1). The empirical basis of White's effect, if there was one, was not obvious despite the considerable effort that we had spent trying to come up with an empirical rationale. This failure underscored the implication of the colorimetric functions: For all but a few simple stimuli, intuition about the empirical experience underlying perception quickly runs out of steam. The only way to understand the perceptions generated by most visual stimuli would be to analyze databases that could serve as proxies for the relevant human experience.

To explain lightness/brightness effects in these terms, Yang began to explore a database of natural scenes in black and white (Figure 10.6). As with the explanation of colorimetry functions, we thought that the images produced by such scenes would approximate human experience with the patterns of light intensities we would have depended on to overcome the inevitably uncertain ability of images to indicate their underlying sources. As with Long's color scene database, we assumed that a collection of natural scenes gathered today would fairly represent the relationships between images and natural sources that humans would always have experienced. If the organization of brightness and lightness perceptions in the brain is indeed a result of the success or failure of behavior in response to retinal luminance patterns over the course of evolution and individual experience, then analyzing a database of this sort would be the only way to understand what we actually see in response to the full range of such stimuli.

Determining the frequency of occurrence of different luminance patterns in scenes is not as difficult as it might seem. For example, consider the standard brightness contrast stimulus in the upper panels of Figure 10.7 or the versions of this pattern shown in Chapter 8. The frequency of occurrence of particular values of luminance corresponding to the central patch in either a dark or light surround will have varied greatly depending on the patterns of retinal luminance

Figure 10.6   Examples of natural scenes from a database of black-and-white images used as a proxy for human experience with the frequency of occurrence of luminance patterns on the retina arising from the world in which we act. (Photos are from the database provided by Van Hateren and Van der Schaaf, 1998.)

generated by typical scenes. Nature being what it is, the luminance value of the central patch in a dark surround will, more often than not, have been less than the luminance value of the central patch in a light surround. The reason is that the patch and surround tend to comprise the same material and be in the same illumination, as is apparent in the scene in the middle panel in Figure 10.7 or the several scenes in Figure 10.6. This bias of the luminance of the central patch toward the luminance of the surround is what humans will always have experienced, and the magnitude of the bias indicated by the frequency of occurrence of the luminance relationships in natural scenes will have guided successful behavior. It follows that this information should be instantiated in visual system circuitry during the course of evolution and individual development, and expressed in the perceptual organization of lightness/brightness.

This universal experience should generate lightness or brightness perceptions in response to a given target luminance according to the

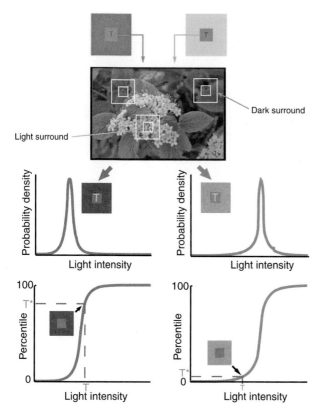

Figure 10.7   Predicting the apparent lightness/brightness of targets in different contexts based on human experience with the retinal luminance patterns generated by natural scenes. The middle panel shows how the stimulus configurations in the upper panels can be used as templates to repeatedly sample scenes such as those in Figure 10.6. In this way, the frequency of occurrence of luminance values for the target (T) in different contexts—such as a light surround versus a dark surround—can be determined. The graphs in the lower panels show that different perceptions of lightness or brightness elicited by the identical luminance values of the targets in the upper panels (the standard brightness contrast effect) are predicted by the different frequencies of occurrence of the same target luminance value in the accumulated experience of humans with dark and light surrounds (expressed as percentiles). See text for further explanation. (After Yang and Purves, 2004. Copyright 2004 National Academy of Sciences, U.S.A.)

relative rank of that luminance among all the possible target luminance values experienced in the context in which it occurred (see the lower panels in Figure 10.7). Despite the directly unknowable real-world sources of the luminance values in retinal images, the relative rank of luminance or some other physical characteristic of light resolves the inverse problem in proportion to the degree that trial-and-error experience that has been incorporated into visual system circuitry. In the case of the standard stimulus in the upper panel, this cumulative human experience would cause the *same* target luminance value in a dark surround to appear brighter or lighter than in a light surround. Using this general approach, Yang showed that the frequency of occurrence of luminance relationships extracted from natural scenes with templates configured in the form of other stimuli predicted a variety of lightness/brightness effects, including White's effect.

A difficulty many people have understanding this approach is how an analysis of scenes serves as a proxy for accumulated human experience. It is relatively easy to grasp the idea that feedback from the success or failure of behavior drives the evolution and development of visual circuitry, and that the connectivity of this circuitry ultimately shapes the organization of perceptual space to reflect the evidence accumulated by trial and error. But where in the analyses of scenes is a representation of the interactions with the world that underlie this scheme? Although not immediately obvious, interactions with the world are represented in the frequency of occurrence of retinal image features generated by the scenes in a database. For example, applying the template in Figure 10.7 millions of times to hundreds of scenes shows how often we would have experienced a particular pattern of light and dark in nature. To be successful, the behavior executed in response to the pattern on the retina would need to have tracked these frequencies of occurrence, which can therefore stand for the lessons that would have been learned by actually operating in the world.

A related point worth emphasizing is that determining the perceptual space of lightness, brightness, or color in this way maintains the relative similarities and differences that objects in the world actually possess. To be biologically useful—for behavior to be successful—surfaces or light sources that are physically the same must look the same, and surfaces or light sources that are different must look different. The construction of visual circuitry according to the frequency of occurrence of light patterns inevitably shapes perception to accomplish this. To appreciate this point, consider the way observers perceive the geometrical attributes of objects, the topic of the next chapter. In relating perceived geometry to physical space, the need to generate perceptions that accord with the physical arrangement of objects is obvious. If geometrical relationships were not preserved, perceptions could not generate motor behaviors that actually worked. To be useful, the perceptions of lightness/brightness elicited by luminance values and the colors elicited by spectral distributions must also be ordered in perceptual space according to the physical similarities and differences among objects and conditions in the world, however discrepant perceptions may be when compared to physical measurements. This pervasive parallelism of perception and physical reality based on operational success in a world that we can't know directly makes it difficult to appreciate that what we see is not the world as it is, and that all perceptions are equally illusory. Indeed, our visual system does this job so well that most people are convinced that what we see is "reality."

Finally, despite the successful prediction of a number of specific perceptions and psychophysical functions pertinent to lightness, brightness, and color in wholly empirical terms, extending this approach to predict more complicated perceptions of these qualities will not be easy. As already noted, the perceptual qualities of lightness/brightness, hue, and saturation are entangled: The visual circuitry that generates lightness/brightness is affected by the circuitry that generates hue, the circuitry that generates hue is affected by the circuitry that generates saturation, and so on. This entanglement occurs because interactions among the various sensory circuits, whether in vision, within some other modality, or among modalities, generate behavior that has a better chance of success. A simple example is the interaction between what we see and what we hear: When we hear a sound, we tend to look in that direction, with improved behavior as a result. It is possible to partially disentangle some of these interactions, as we did by using databases that focused on only one type of information (such as black-and-white scenes or scenes that include color spectra). But the interplay among sensory qualities within a modality (such as how hue affects brightness), among sensory modalities (such as how what we hear influences what we see), and even among sensory and nonsensory functions (such as how what we think, feel, or remember influences what we see and hear) means that the definition of the perceptual space of any quality can be complete only when all the brain systems that influence that space are taken into account. The highly interactive organization of brains is enormously useful in generating behavior that has the best chance of success. Nonetheless, the biologically necessary entanglements among perceptual spaces will make understanding more complex perceptions in empirical terms increasingly difficult.

11

Perceiving geometry

In general, the complex discrepancies between the physical world and the perceptual spaces that define lightness, brightness, and color can be understood in terms of an empirical strategy that has evolved to circumvent the inverse problem. If this is the way vision operates, the same scheme should also explain perceptual phenomena pertinent to all the other visual qualities we see, preferably in quantitative terms. One of the most important of these additional qualities is the perception of geometry, the way we see spatial intervals, angles, shapes, and distances. Obviously, perceiving these fundamental aspects of physical geometry in a manner that enables appropriate behavior is critical. And here again biological vision must deal with the direct unknowability of the world by means of light from the environment falling on the retina.

In some ways, understanding the inverse problem and its consequences in the context of geometry is easier than understanding the similar quandary in the domains of lightness, brightness, and color. For instance, it's easy to appreciate that objects of different sizes with different orientations and at different distances from observers can all produce the same retinal image (Figure 11.1). It was these facts about projective geometry that Berkeley had used as a basis for his arguments about vision in the early eighteenth century (see Figure 7.8).

The perception of geometry offers an abundance of perceptual weirdness. Numerous discrepancies between measurements made with rulers or protractors and the corresponding perceptions have been described, providing plenty of challenges for any explanation of

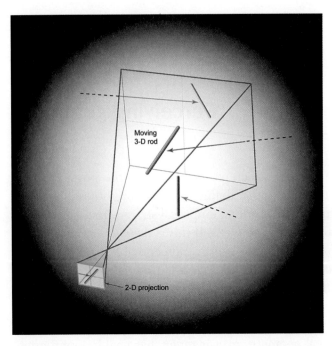

Figure 11.1   The inverse problem as it pertains to geometry. The same projec-tion on the retina can be generated by many physically different objects in three-dimensional space. This inherent uncertainty applies to the geometry of any retinal image. (Courtesy of Bill Wojtach)

this aspect of vision (Figure 11.2). If the circuitry of the visual brain is determined by responses to projections on the retina whose sources can't be known directly, these phenomena should be also accounted for by the frequency of occurrence of different geometrical configu-rations in retinal images generated by the geometry of the world.

Catherine Qing Howe, another superb product of the Chinese educational system, was the person who met this challenge head on. Qing had been a graduate student for about three years in the Department of Neurobiology at Duke when she came to see me one day in 2000 in obvious distress. She had begun a doctoral dissertation on ion channels with a molecular biologist in the department but had become increasingly disenchanted with her project, her mentor, and molecular biology as a way to pursue her interests. She indicated that she wanted to start over by joining my lab. Such shifts in direc-tion were unusual, and switching from the molecular biology of ion

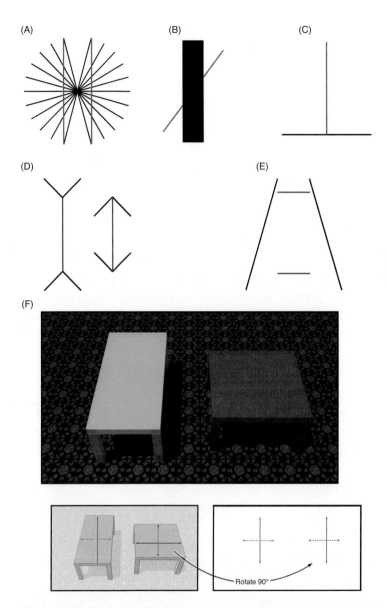

Figure 11.2   Discrepancies between measured geometry and what we see. A) Two parallel lines appear bowed when presented on a background of converging lines. B) Segments of a continuous line obscured by a bar appear to be vertically displaced. C) The same line appears longer when it is vertical than horizontal. D) A line terminated by arrowheads looks longer than the same line terminated by arrow tails. E) The same line appears longer when presented in the narrower region of the space between diverging lines. F) The shape and size of the surfaces with the same dimensions (see the key below) can look very different. (After Purves and Lotto, 2003)

channels to perception was about as radical a change as one could imagine in neuroscience. But she eventually convinced me that this was really what she wanted to do.

At age 10, Qing had been chosen by the Chinese Academy of Sciences as one of 30 intellectually gifted children to receive an individualized curriculum in the Beijing Middle School system. She had entered Peking Union Medical College at age 15, the youngest student they had ever taken. There she had became increasingly interested in cognition and behavior, and had decided to pursue these topics in the context of psychiatry. Her idea (much as mine had been as a first-year medical student with similar intentions) was that the best way to understand these issues was through molecular pharmacology. During her last year in med school, she was elected to represent her class in an exchange program with the University of California San Francisco Medical School. The experience convinced her to immigrate to the United States, and after receiving her medical degree in 1997, she entered the neurobiology graduate program at Duke.

Qing quickly recognized that if we were going to explain the phenomena in Figure 11.2 in empirical terms, a first step would be to determine the frequency of occurrence of geometrical images projected onto the retina, much as we had begun to assess the frequency of luminance and spectral distributions in images generated by sources in the natural world. Happily, there was a relatively easy way to acquire this information. Laser range scanning is a technique routinely used to monitor the geometrical conformance of an architectural plan to the progress of the building under construction. The device commonly used provides accurate measurements of the distances from the origin of the scanner's laser beam to all the points (pixels) in a digitized scene. By setting the height of the scanner at the average eye level of human observers, we could evaluate the projected geometry of retinal images routinely generated by real-world objects. The summer after Qing joined the lab, we could be seen lugging this machine and its accessories around the Duke campus to scan a variety of scenes (Figure 11.3).

A database of this sort has some serious limitations. The range of distances analyzed was restricted to a few hundred meters, and

A

Figure 11.3 Determining the relationships between images and the physical geometry of objects in the world. A) A laser range scanner. A mirror inside the rotating head directs a laser beam over a scene; the signal reflected back from object surfaces is detected by a photodiode, which, in turn, produces an electrical signal. A quartz clock determines the tiny interval between the transmitted pulse and the signal from the returning beam; based on the speed of light, a microcomputer calculates the distance from the image plane of the scanner to each point on object surfaces. B) Ordinary digital images of a natural scene, and an outdoor scene that contains some human artifacts. C) The corresponding range images the scanner acquired. Color-coding indicates the distance from the image plane of the scanner to each point in the scenes. (From Howe and Purves, 2005. Copyright 2005 Springer Science+Business Media. With kind permission of Springer Science+Business Media.)

the scenes were acquired in a particular locale (the Duke campus) and in a particular season (summer). Moreover, when human observers look at the world, they don't do so in the systematic manner of a laser scanner: They fixate on objects and parts of objects that contain information particularly pertinent to subsequent behavior (such as other people, faces, and object edges). Nonetheless, the database provided a reasonable approximation of the prevalence of different two-dimensional geometries in images generated by 3-D sources in the world. To test the hypothesis that the geometry we see is determined by accumulated information about the frequency of occurrence of various geometrical projections, Qing sampled thousands of images in the database with templates configured in the same form as a stimulus pattern of interest (for example, the geometrical stimuli in Figure 11.2). By sampling a large data set pertinent to a particular stimulus, she could ask whether the way people actually see the geometry of the retinal image is determined empirically.

To understand this approach, consider the perceived length of a line compared to its actual length in the retinal image. In human experience, the length of a line on the retina will have been generated by lines associated with objects in the real world that have many different actual lengths, at different distances from the observer, and in different orientations (see Figure 11.1). As a result, it would be of no use to perceive the length of the line in the retinal image as such, just as it would be of little or no use to see luminance or the distribution of spectral energy as such. To deal successfully with the geometry of objects in the world, it would make far more sense to generate perceived lengths empirically. The length seen would be determined by the frequency of occurrence of any particular length in the retinal image relative to all the projected lengths experienced by human observers in the same orientation. In keeping with the argument in Chapter 10, this relationship discovered by feedback from trial-and-error behavior would have shaped the perceptual space for line lengths. As a result, the lengths seen would always differ from the geometrical scale of lengths measured with a ruler (Figure 11.4).

For instance, if in past human experience 25% of the lines in retinal stimuli generated by objects in the world have been shorter than or equal to the length of the stimulus line in question, the rank of that projected length on an empirical scale would be the 25th percentile. If the length of another line stimulus had a rank of, say, the 30th percentile,

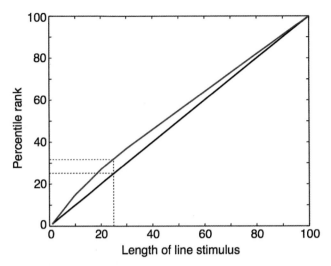

Figure 11.4 An empirical scale based on the frequency of occurrence of line lengths in retinal images (red), compared to a scale of lengths measured with a ruler (black). Because any given line length (25 units in the example here) has a different rank on these two scales (dotted lines), if what we see is determined empirically, there will always be a discrepancy between the perceived length of a line and its measured length in the retinal stimulus or in the world. (After Howe and Purves, 2005. Copyright 2005 Springer Science+Business Media. With kind permission of Springer Science+Business Media.)

then the stimulus at the 25th percentile should appear shorter, to a degree determined by the two ranks. The consequence of this way of perceiving line lengths—or spatial intervals, generally—is routine discrepancies between measurements of lines with rulers and the subjective "metrics" that characterize perception. As with lightness, brightness, or color, seeing visual qualities according to their empirical ranks maintains the relative similarities and differences among physical objects that are pertinent to successful behavior despite the direct unknowability of geometry in the world. The strategy works so well that we imagine that the geometry we see represents the actual geometry of objects, leading to the erroneous idea that the demonstrations in Figure 11.2 are "illusions." In fact, they are signatures of the way the visual system contends with this aspect of the inverse problem.

To appreciate how this framework can explain the anomalies in Figure 11.2, take the variation in the perceived length of a line as its orientation is varied (Figure 11.5A). As investigators have repeatedly shown over the last 150 years, a line looks longer when presented vertically than horizontally; oddly, however, the maximum length is seen

when the stimulus line is oriented about 30° from vertical
(Figure 11.5B). In the empirical framework of perceived geometry
that Qing pursued in her thesis, the apparent length elicited by a line
of any given projected length and orientation on the retina should be
predicted by the rank of the line on an empirical scale determined by
the frequency of occurrence of that projection (see Figure 11.4).
Again, the reason is that this information would have been incorporated in visual circuitry to maximize the success of behavior directed
toward sources whose lengths cannot be known directly.

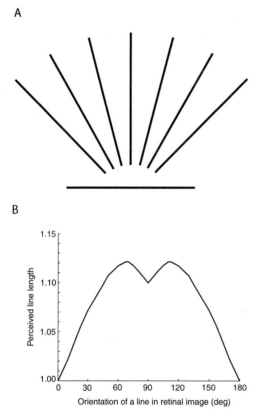

Figure 11.5 Differences in the apparent length of the same line as a function of
its orientation. A) The horizontal line in the figure looks somewhat shorter than the
vertical or oblique lines, despite the fact that all the lines are identical (see also
Figure 11.2C). B) The apparent length of a line reported by subjects as a function
of orientation, expressed in terms of the angle between a given line and the horizontal axis. The maximum length seen by observers occurs when the line is oriented approximately 30° from vertical, at which point it appears about 10%–15%
longer than the minimum length seen when the orientation of the stimulus is horizontal. (After Howe and Purves, 2005. Copyright 2005 Springer Science+Business
Media. With kind permission of Springer Science+Business Media.)

Qing tested the merits of this explanation by assessing the frequency of straightline projections generated by the objects in the laser-scanned scenes (Figure 11.6). By extracting all the projected

A   B

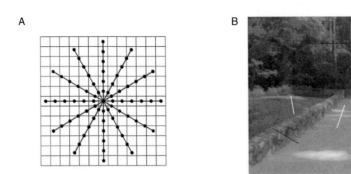

Figure 11.6   The frequency of occurrence of lines in different orientations on an image plane, determined by analyzing a database of laser-scanned scenes. A) The pixels in part of an image from the database are represented diagrammatically by the grid squares; the connected black dots indicate a series of templates used to determine the frequency of occurrence of straight line projections at different orientations in images arising from straight lines in the 3-D world (note that the definition of straight lines is geometrical and not dependent on visible edges alone). B) Examples of straight-line templates overlaid on a typical image, for which the corresponding distance and direction of each pixel are known (see Figure 11.3). White templates indicate sets of points that correspond to straight lines in the world, and red templates indicate sets that do not. By repeating such sampling millions of times in different images, the frequency of occurrence of straight line projections of different lengths in different orientations on the retina can be tallied. (After Howe and Purves, 2005. Copyright 2005 Springer Science+Business Media. With kind permission of Springer Science+Business Media.)

straight lines from the database that corresponded to geometrical straight lines on surfaces in the 3-D world, she could compile the frequency of occurrence of projected lines at different orientations. In effect, this analysis represents human experience with lines of different lengths and orientations in retinal images. For a line of any particular length and orientation on the retina, some percentage of projected lines in that orientation will have been shorter than the line in question, and some percentage will have been longer (Figure 11.7). In an empirical framework, this accumulated experience will have shaped the perceptual space of line length and thus the length of the lines that observers see. If this idea is right, the probability distributions in Figure 11.7 should predict the puzzling psychophysical results in Figure 11.5B.

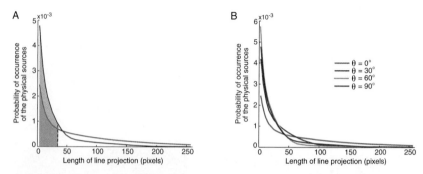

Figure 11.7 Experience with straight-line projections in different orientations generated by the geometry of the world. A) The relative occurrence of projected line lengths in vertical (red) and horizontal (blue) orientations. The area under the two curves indicates accumulated human experience with projected lines of any given length in these two orientations. For a vertical line of a particular length (dashed line), humans will have experienced relatively more lines that are shorter than this length (area under the red curve to the left of the dashed line) compared to experience with a horizontal line of the same projected length (area under the blue curve). B) Probabilities of occurrence of lines at other orientations pertinent to explaining the psychophysical function in Figure 11.5B. See text for further explanation. (After Howe and Purves, 2005. Copyright 2005 Springer Science+Business Media. With kind permission of Springer Science+Business Media.)

In thinking about this explanation of apparent length, recall the basic problem. Although a ruler accurately measures physical length, the visual system cannot do so because of the inverse problem (see Figure 11.1). (Note, incidentally, that the perceived length of a ruler is no more veridical than any other length we see; the markings on it vary in separation according to the orientation of its projected length on the retina, as indicated in Figure 11.5.) However, if the visual system ordered the perception of projected lengths according to feedback from accumulated experience, the inverse problem could be circumvented: Lengths on the retina would be operationally associated with behavior directed at physical lengths according to the frequency of occurrence of projected lengths. Although visually guided behavior on this basis is statistically determined, enough experience over evolutionary and individual time would make actions in the world increasingly efficient.

If this scheme is indeed being used to generate biologically useful perceptions, the puzzling variation in the apparent length of a line as a function of its orientation (see Figure 11.5B) should be predicted by the frequency of occurrence (the empirical rank) of projected line lengths as orientation changes. When Qing used the data in Figure 11.7 to predict how the same line at different orientations would be seen on this basis, she found that the prediction given by the empirical rank of lines in different orientations matched the McDonald's arches–like function that describes the lengths people actually see (Figure 11.8). We were all impressed with this result; it was hard to imagine that this odd psychophysical function could be explained in any other way.

A. Reported by Subjects

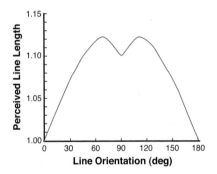

B. Predicted Empirically

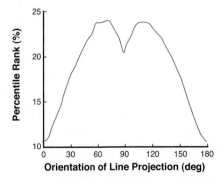

Figure 11.8 Comparing the line lengths reported by subjects (A; taken from Figure 11.5B), and the perception of line length predicted from the frequency of occurrence of differently oriented straight lines in the retinal images (B). The prediction is for a particular projected length, but the same general shape is apparent for any length. (After Howe and Purves, 2005. Copyright 2005 Springer Science+Business Media. With kind permission of Springer Science+Business Media.)

Why, then, are there more sources in the world—and, thus, more projected images—of relatively short vertical lines compared to horizontal lines, and why are there even more sources that project as relatively short lines at 20°–30° away from vertical (see Figure 11.7)? Straight lines in the physical world are typically components of planes, a statement that may seem odd because we are very much aware of lines at contrast boundaries (for example, the edges of things). But whereas explicit lines generated by contrast obviously provide useful information about the edges of objects, object surfaces are by far the more frequent source of the geometrical straight lines that we experience. Thus, when considering the physical sources of straight lines that project onto the retina in different orientations, the most pertinent variable is the extension of relatively flat surfaces. Horizontal line projections in the retinal image are typically generated by the extension of planar surfaces in the horizontal and depth axes, whereas vertical lines are typically generated by the extension of surfaces in the vertical axis and the depth axis (Figure 11.9A). The generation of line projections from the extension of surfaces in depth is inherently limited because the depth axis is perpendicular to the image plane; lines in this plane are thus foreshortened by perspective (that is, they generate shorter lines on the retina). A quick inspection of the world makes clear that the extension of surfaces in the vertical axis is also limited by gravity, thus increasing the prevalence of shorter vertical projected lines (overcoming gravity takes work, so objects, natural or otherwise, tend to be no taller than they have to be). Because this limitation does not restrict the generation of horizontal line projections arising from the ground plane, humans experience more short vertical line projections than short horizontal ones (see Figure 11.7). As a result, a vertical line on the retina will always have a higher empirical rank than a horizontal line of the same length (that is, more lines shorter than the vertical line will have been projected on the retina, giving the vertical line a relatively higher rank in the perceptual space of apparent lengths than a horizontal line of the same length).

A different real-world bias accounts for the fact that there are more short line projections 20°–30° away from vertical than dead-on vertical, giving these oblique lines the highest empirical rank and thus the greatest apparent length (see Figures 11.8 and 11.5B).

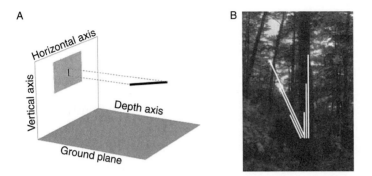

Figure 11.9 Physical bases for the biased projection of lines in different orientations. A) The projection of long vertical lines is limited by foreshortening and by the relative paucity of tall vertical objects in the world. As a result, humans will always have been exposed to more short vertical line projections than short horizontal ones. B) Image of a natural scene with superimposed vertical lines and oblique lines about 25° from the vertical axis. Despite the overall paucity of long vertical line projections compared to horizontal ones, longer vertical line projections are more likely than oblique ones because of the relative abundance of vertical compared to oblique surfaces in the world. See text for further explanation. (After Howe and Purves, 2005. Copyright 2005 Springer Science+Business Media. With kind permission of Springer Science+Business Media.)

To understand this peculiarity, consider vertical and oblique lines of the same lengths superimposed on natural scenes. As illustrated in Figure 11.9B, natural surfaces provide more sources of relatively long vertical lines than oblique lines, as indicated by the data for lines at 60° in Figure 11.7B. The reason is that natural surfaces such as tree trunks tend to extend vertically instead of obliquely because of greater mechanical efficiency. As a result, relatively long linear projections somewhat away from vertical are less frequent than equally long vertical projections. Shorter projected lines oriented about 25° away from vertical are, therefore, even more common than shorter vertical lines (although shorter vertical lines are more common than horizontal ones; see above). The upshot is that their empirical rank is somewhat greater than the rank of vertical lines, causing the McDonalds-like arches in Figures 11.5B and 11.8B.

Another challenge in rationalizing geometrical percepts on an empirical basis is perception of the apparent angle made by two lines that meet (either explicitly or implicitly) at a point. As with the apparent length of lines, an intuitive expectation about the perception of angles is that this basic feature of geometry should scale directly with the size of angles measured with a protractor. However, this is not what people see. It has long been known that observers tend to over-estimate the magnitude of acute angles and underestimate obtuse ones by a few degrees (Figure 11.10A). The anomalous perception of angles is easiest to appreciate—and most often demonstrated—in terms of a related series of geometrical stimuli that involve intersect-ing lines in various configurations. The simplest of these is the so-called tilt illusion, in which a vertical line in the context of an obliquely oriented line appears to be slightly rotated away from verti-cal in the direction opposite the orientation of the oblique "inducing line" (Figure 11.10B). The direction of the perceived deviation of the vertical line is consistent with the perceptual enlargement of the acute angles in the stimulus and/or a reduction of the obtuse angles. This relatively small effect is enhanced in the Zöllner illusion, a more elaborate version of the tilt effect, achieved by iterating the basic fea-tures of the tilt stimulus (Figure 11.10C). The several parallel vertical lines in this presentation appear to be tilted away from each other, again in directions opposite the oblique orientation of the contextual line segments (see also 11.2A, which depends on this same effect). The challenge for an empirical interpretation of vision is whether accumulated experience with retinal images and their sources in the world can also explain the odd way we see angles.

On empirical grounds, these differences between measured and perceived angles are expected. Just as the inverse problem makes the source of a projected line unknowable (see Figure 11.1), an angle on the retina could arise from any real-world angle. Perceiving angles on an empirical basis would allow observers to contend with this inevitable ambiguity. Understanding angles in these terms depended on much the same approach as understanding the perception of interval lengths. The frequency of occurrence of angle projections generated by the geometry of the world could be determined from laser range images, and Qing's supposition was that this information would have determined the way we see angles and should thus explain perceptual anomalies such as those in Figure 11.10. A first step was to identify regions of the laser-scanned scenes in the data-base of real-world geometry that contained a valid physical source of one of the two lines that form an angle (the black reference line in

(A)

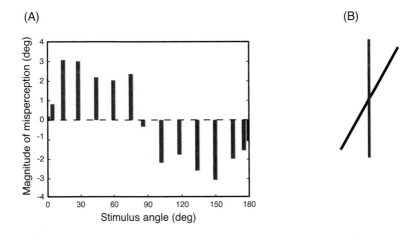

(B)

(C)

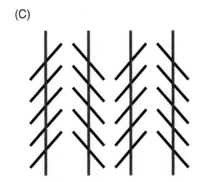

Figure 11.10 Discrepancies between measured angles and their perception. A) Psychophysical results showing that acute angles are slightly overestimated and obtuse ones underestimated. B) The tilt illusion. A vertically oriented test line (red) appears tilted slightly counterclockwise in the context of an oblique inducing line (black). C) The Zöllner illusion. The vertical test lines (red) appear more impressively tilted in directions opposite the orientations of the contextual lines (black) when the components of the tilt effect are repeated. (Data in A are from Nundy, et al., 2000 [Copyright 2000 National Academy of Sciences, U.S.A.]; B, C after Howe and Purves, 2005 [Copyright 2005 Springer Science+Business Media. With kind permission of Springer Science+Business Media.])

Figure 11.11A). After a valid reference line had been found, the occurrence of a valid second line forming an angle with it could be determined by overlaying a second straight-line template in different

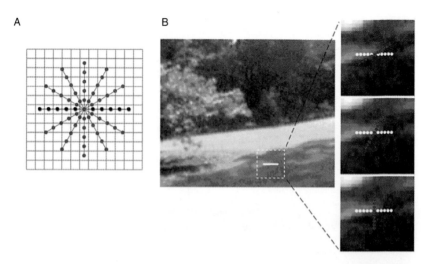

Figure 11.11 Determining the frequency of occurrence of angles generated by the geometry of the world. A) As in Figure 11.6, the pixels in an image are represented by grid squares. The black dots indicate a reference line template and the red dots indicate additional templates for sampling a second line making different angles with the reference line. B) The white line overlaid on the images indicates valid reference lines. Blowups of the boxed area show examples of the second template (red) that was overlaid on the same area of the image to sample for the presence of a second straight line making a valid angle (in each of the cases shown, the second template is also valid). (After Howe and Purves, 2005. Copyright 2005 Springer Science+Business Media. With kind permission of Springer Science+Business Media.)

orientations on the image (the red lines in Figure 11.11). By systematically repeating this procedure, the relative frequency of occurrence of different projected angles on the retina could be tallied.

Figure 11.12 shows the frequency of occurrence of projected angles projected on the retina. Regardless of the orientation of the reference line (indicated by the black line in the icons under the graphs) or the type of real-world scene from which the statistics are derived, the likelihood of angle projections is always least for 90° and greatest for angles that approach 0° or 180°. In other words, the chances of finding real-world sources of an angle decrease as the two lines become increasingly perpendicular.

The cause of the bias evident in these statistical observations can, like the biased projection of line lengths in different orientations, be understood by considering the provenance of straight lines in the physical world. As with single lines, intersecting straight lines in the

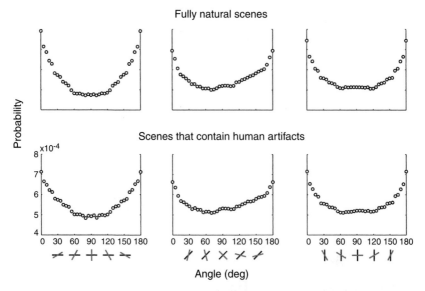

Figure 11.12   Frequency of occurrence of angles projected onto the retina by the geometry of world. The three columns represent the frequency of occurrence of projected angles based on different orientations of the sampling template (indicated by the icons below the graphs). The upper row shows the results obtained from fully natural scenes, and the lower row shows the results from environments that contained some (or mostly) human artifacts (see Figure 11.3 for examples). (After Howe and Purves, 2005. Copyright 2005 Springer Science+Business Media. With kind permission of Springer Science+Business Media.)

world are typically components of planar surfaces. Accordingly, a region of a planar surface that contains two physical lines whose projections intersect at 90° will, on average, be larger than a surface that includes the source of two lines of the same length but intersecting at any other angle (Figure 11.13). Because larger surfaces include smaller ones, smaller planar surfaces in the world are inevitably more frequent than the larger ones. Thus, other things being equal, physical sources capable of projecting angles at or near 90° are less prevalent than the sources of any other angles.

If perceptions of angle magnitude are generated on the basis of past experience with the frequency of occurrence of projected angles, the angles seen should accord with their relative empirical rank of angle magnitude determined in this way. Figure 11.14A shows how the frequency of occurrence of angle projections derived from the physical geometry of the world would be expected to influence the

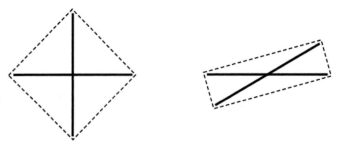

Figure 11.13   The physical source of two lines of the same length intersecting at or near 90° must be a larger planar surface (dashed line) than the source of the same two lines, making larger or smaller angles. This geometrical fact explains the lower probability of 90° projections in Figure 11.12 compared to other angles. See text for further explanation. (From Howe and Purves, 2005. Copyright 2005 Springer Science+Business Media. With kind permission of Springer Science+Business Media.)

perceptual space for angles. The empirical rank of any angle between 0° and 90° is shifted slightly in the direction of 180° compared to actual geometrical measurements, and the opposite is true for any angle between 90° and 180°. Accumulated experience with the relative frequency of angle projections on the retina generated by the geometry of the world thus predicts the psychophysical results shown in Figure 11.10 (Figure 11.14B, C).

Not surprisingly, many theories have been proposed over the last century and a half to explain the anomalous perception of line lengths, angles, and related issues such as perceived object size. Attempts to rationalize these discrepancies in the nineteenth century included asymmetries in the anatomy of the eye, the ergonomics of eye movements, and cognitive compensation for the foreshortening in perspective. More recent investigators have supposed that the anomalous perceptions of geometry arise from inferences based on personal experience. For example, British psychologist Richard Gregory argued that the different line lengths seen in response to the stimulus in Figure 11.2D (the Müller–Lyer illusion) are a result of interpreting the arrow-tails and arrowheads in the stimuli as concave and convex corners, respectively. The anomalous perception of line length was taken to follow from the different distances implied by such real-world corners, with the assumption that convex corners implied by the arrowheads would be nearer to the observer than concave corners

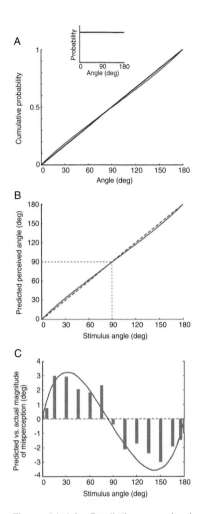

Figure 11.14 Predicting perceived angles based on the frequency of occurrence of images generated by real-world sources. A) The red curve shows the frequency of occurrence (expressed as cumulative probability) of projected angles derived from the data in Figure 11.12. For comparison, the black line shows the frequency of occurrence that would be generated if the probability of any projected angle was the same (see inset). B) The red curve shows the perception of angle magnitude predicted from the information in (A) (the dashed diagonal indicates the angle projected on the retina and the thin dashed line indicates 90°). C) The predictions in (B) compared to psychophysical measurements of angle perception taken from Figure 11.10A. (After Howe and Purves, 2005. Copyright 2005 Springer Science+Business Media. With kind permission of Springer Science+Business Media.)

implied by arrow tails. Although such explanations are empirical, analysis of real-world geometry often contradicts intuitions that seem obvious. For example, laser scanning of buildings and rooms shows that there is no significant difference in the distance from observers of convex and concave corners. As in lightness, brightness, and color, intuitions are a poor foundation for understanding perception.

Contemporary explanations of these effects have more often turned to the receptive field properties of visual neurons (see Chapters 1 and 7), suggesting, for example, that lateral inhibitory effects among orientation-selective cells in the visual cortex could underlie the anomalous percepts illustrated in Figure 11.10. In this interpretation, the perception of an angle might differ from its actual geometry because orientation-selective neurons coactivated by the two arms of the angle inhibit each other. The supposed effect would be to shift the distribution of cortical activity toward neurons whose respective selectivity would be farther apart than normal, thus explaining the perceptual overestimation of acute angles (some other interaction would be needed to explain the underestimation of obtuse angles). As with some attempts to explain lightness/brightness effects described in Chapter 8, this approach implies that perceptual discrepancies are a side effect of other goals in visual processing.

In short, a lot of explanations have been proposed for the anomalies evident in the perception of geometrical forms, and some remain in play. The advantage of the empirical explanation provided by Qing is that it covers the full range of geometrical phenomena to be accounted for (only a few examples have been described here) and provides a strong biological rationale: contending with the inverse problem as it pertains to the geometry of retinal projections. It helps that this way of accounting for geometrical percepts also accords with the empirical explanation of many aspects of lightness, brightness, and color.

Nonetheless, many people have difficulty understanding how this way of seeing apparent lengths, angles, or other aspects of geometry could possibly be helpful in guiding behavior. Surely seeing these features for what they are in the image or the world would be the most logical way to relate perception to behavior, even though the inverse problem precludes direct specification of physical sources. But to reiterate, the seeing image features would be not be useful, and the idea that they should be misses the nature of the problem that vision must solve. Only by encoding the empirical results of trial-and-error discovery from successful behavior is it possible to deal with the world effectively through the senses. Seeing geometry the way we do is a reflection of this process.

12 Perceiving motion

Another key perceptual quality to consider in exploring how the visual brain works is motion. Although Zhiyong Yang had struggled early on to make some sense of motion in empirical terms, we had put this issue aside while focusing on brightness, color, and geometry as perceptual domains that would be more tractable in trying to understand how brains use sense information that can't specify the physical world directly. The main reason for this reticence was that we had no idea how to determine object motion in three-dimensional space, the information we needed to test the idea that perceptions of motion arise empirically. A less straightforward concern was the conceptual difficulty that an empirical explanation of motion entails. It was hard to imagine that successful behavior in response to moving objects could possibly be based on the frequency of occurrence of retinal stimulus sequences. Most vision scientists take it for granted that the brain generates perceptions of motion and visually guided actions using the features in retinal images to compute object motion "online." Even for us, the idea that the perception of motion and the complex actions involved in a motor response such as catching a ball could be based on trial-and-error behavior seemed bizarre.

In physical terms, motion refers to the speed and direction of objects in a three-dimensional frame of reference and follows Newton's laws. In psychophysical terms, however, seeing object speeds and directions is defined subjectively. For example, we don't see the motion of a clock's hour hand or a bullet in flight, even though both objects move at physical rates that are easily measured. The range of object speeds projected onto the retina that humans have evolved to see as motion is from roughly 0.1° to 175° of visual angle per second, a degree of visual angle being 1/360 of an imaginary circle around the head (about the width of a thumbnail at arm's length). Below this range, objects appear to be standing still. As speeds approach the upper end of the range, moving objects begin to blur and then, as with the bullet, become invisible.

One of the things that had puzzled natural philosophers and neuroscientists thinking about motion is the obvious way that context affects motion perception. Depending on the circumstances, the same speed and direction projected onto the retina can elicit very different perceptions. Although such phenomena were often treated as special cases or were simply ignored, if what we had been saying about vision was true, then the perception of motion (including these anomalies) should have the same empirical basis as lightness, brightness, color, and perceptions of geometry. All the motions we see should be understandable in terms of the biological need to contend with the inverse problem as it applies to moving objects.

How the inverse problem pertains to seeing motion is easy enough to understand. For obvious reasons, observers must respond correctly to the real-world speeds and directions of objects, and these responses are certainly initiated by the speeds and directions of objects that determine stimulus sequences projected onto the retina. But as illustrated in Figure 12.1, when objects in three-dimensional space are projected onto a two-dimensional (2-D) surface, speed and direction are conflated in the resulting images. As a result, the sequence of actual positions in 3-D space that define motion in physical terms is always uncertainly represented in the sequence of retinal positions that moving objects generate.

If contending with this problem depended on the empirical framework that we had used to rationalize other visual qualities, the perceptions of motion elicited by image sequences should accord with—and be predicted by—the frequency of occurrence of the retinal image sequences generated by moving objects in the world. However, testing this idea was not so easy. The most formidable obstacle was acquiring a database of the 2-D projections arising from the speeds and directions of objects moving in 3-D space. Although data relating retinal projections to real-world geometry had been relatively easy to obtain for static scenes using laser range scanning (see Chapter 11), we had no technical way to get the information about

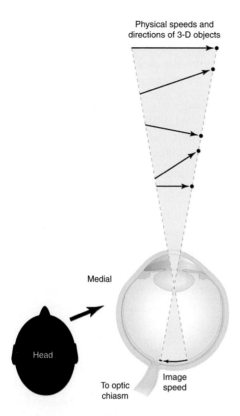

Figure 12.1   The inverse optics problem as it pertains to the speed and direction of moving objects projected onto the retina. Objects (black dots) at different distances moving in different trajectories (arrows) with different speeds can all generate the same 2-D image sequence on the retina (the diagram shows a horizontal section through the right eye). Therefore, speeds and directions in retinal images cannot specify the speeds and directions of the real-world objects that observers must deal with. (From Wojtach, et al., 2008. Copyright 2008 National Academy of Sciences, U.S.A.)

the direction and speed of objects moving in space that we needed to determine the frequency of occurrence of speeds and directions on the retina.

Nevertheless, we could approximate human experience with object motion using a virtual world in which moving objects were projected onto an image plane (a stand-in for the retina) in a computer simulation (Figure 12.2). Although grossly simplified with respect to all the empirical factors that influence natural motion, this surrogate for experience with moving objects was not unreasonable. Most important, it accurately represented *perspective*, the major determinant of the difference between object motion in 3-D space and the speeds and directions projected onto the 2-D retinal surface (see Figure 12.1). By sampling the image plane in the virtual environment, we could tally up the frequency of occurrence of the projected speeds and directions arising from 3-D sources whose speeds and directions were known. We could then use this approximation of motion experience to predict the perceived speeds and directions that should be seen in response to various motion stimuli if motion perception is generated empirically.

As in every other perceptual domain, there are puzzles in motion perception whose bases have been debated for decades. Two specific examples that people have struggled to explain are the flash-lag effect, which concerns the perception of speed, and the effects of occluding apertures, which concern the perception of direction (recall that speed and direction are the two basic characteristics of perceived motion). Two postdocs, Kyongje Sung and Bill Wojtach, took on the challenge of explaining these effects in empirical terms. Kyongje had earned his Ph.D. at Purdue in 2005 studying how people carry out visual search tasks (looking for a specific object in a scene), and he was the first card-carrying psychophysicist to join the lab. Bill had gotten his Ph.D. at about the same time in philosophy at Duke working on perception; he had come to lab meetings while working on his doctorate and eventually decided to pursue a career that tapped into both philosophy and neuroscience. Wojtach and Sung made a somewhat unlikely scientific pair, but they complemented each other's skills nicely and eventually succeeded in making the case that perceptions of motion are indeed based on the same strategy as other visual qualities, and for the same general reasons.

They first explored the flash-lag effect, a phenomenon that had been studied since the 1920s without any agreement about its cause. When a continuously moving stimulus is presented in physical alignment with an instantaneous flash that marks a point in time and space, the flash is seen as lagging behind the moving stimulus (Figure 12.3A). Moreover, the faster the speed of the stimulus, the greater the lag (Figure 12.3B). The effect is actually one of several related phenomena apparent when people observe stimuli in which a moving object is presented together with an instantaneous marker. For

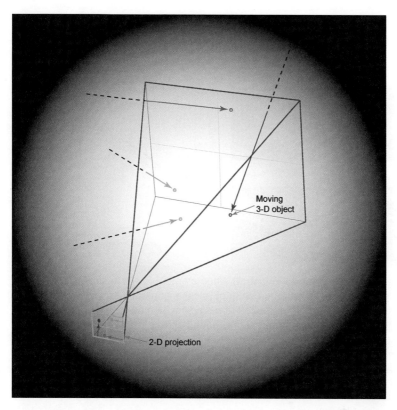

Figure 12.2 Determining the frequency of occurrence of image sequences generated by moving objects in a virtual environment. Diagram of a simulated visual space (red outline) embedded in a larger spherical environment; objects moving randomly in different directions at different speeds enter the visual space and are projected onto an image plane (blue outline). By monitoring the projections, we could determine the frequency of occurrence of the speeds and directions in image sequences generated by objects moving in 3-D space. (After Wojtach, et al., 2008. Copyright 2008 National Academy of Sciences, U.S.A.)

example, if the flash occurs at the start of the trajectory of an object, the flash is seen as displaced in the direction of motion (the so-called Fröhlich effect). And if observers are asked to specify the position of a flash in the presence of nearby motion, the flash is displaced in that direction (the flash-drag effect). People don't ordinarily notice these discrepancies, but they are quite real and reveal a systematic difference between the speed projected on the retina and the speed we see. This difference raises questions about how the perception of speed is related to behavior and why the discrepancies exist in the first place. Despite various proposed explanations, there were no generally accepted answers.

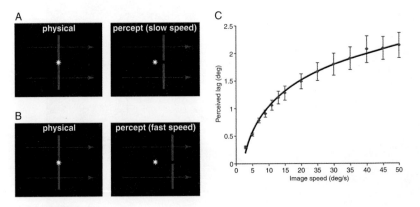

Figure 12.3   The flash-lag effect. A) When a flash of light (indicated by the asterisk) is presented on a screen in physical alignment with moving stimulus (the red bar), the flash is perceived to be lagging behind the position of the moving object. B) The apparent lag increases as the speed of the moving object increases. C) The amount of lag seen as a function of object speed, determined by asking subjects to adjust the position of the flash until it appeared to be in alignment with the moving stimulus. (After Wojtach, et al., 2008. Copyright 2008 National Academy of Sciences, U.S.A.)

In the framework we had been pursuing to explain other visual qualities, the flash-lag and related effects would be signatures of this same empirical strategy applied to the perception of object speed. To test this supposition, Wojtach and Sung asked whether the amount of lag subjects see over a range of speeds is accurately predicted by the relative frequency of occurrence of image sequences arising from 3-D object motion transformed by projection onto the retina (see Figure 12.1). The first step was to determine the relevant psychophysical function by having observers align a flash with the moving bar, while we varied the speed of the moving object over the full range that elicits a measurable flash-lag effect (Figure 12.3C). To test whether this function accords with an empirical explanation of perceived speed, Sung repeatedly sampled the image sequences generated by objects moving in the simulated environment, tallying the frequency of occurrence of the different projected speeds generated by millions of possible sources moving in the simulated 3-D world (see Figure 12.2). In an empirical framework, the speeds that observers see should be given by the relative frequencies of occurrence of projected speeds on the retina, which in turn would have determined how observers see object motion. If the flash-lag effect is indeed a signature of visual motion processing on an empirical basis, the lag that observers report for different stimulus speeds should be accurately predicted by the relative ranks of different image speeds in perceptual space arising from this experience. As shown in Figure 12.4, the psychometric function in Figure 12.3C is closely matched by accumulated experience with the speeds on the retina that moving objects generate.

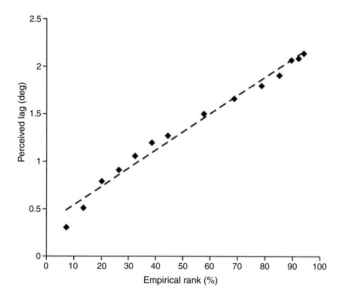

Figure 12.4   Empirical prediction of the flash-lag effect. The graph shows the perceived lag that observers report (see Figure 12.3C) plotted against the empirical rank (relative frequency of occurrence) of the projected image speeds that moving objects generate. The diamonds are the empirically predicted lags; the dashed line indicates perfect correlation between the predicted lags and the amount of lag subjects report. (From Wojtach, et al., 2008. Copyright 2008 National Academy of Sciences, U.S.A.)

This explanation of perceived speed is similar to the accounts given for lightness, brightness, color, and geometry. As in the subjective organization of these other perceptual qualities, contending with the inverse problem requires that the speeds we see be organized in perceptual space according to the frequency of occurrence of projected speeds on the retina (because this information permits successful behavior). Since the flash is not moving, it will always have a lower empirical rank than a moving stimulus (such as the moving bar in Figure 12.3) and therefore should appear to lag behind the position of any moving object projected on the retina. And because increases in image speed correspond to a higher relative rank in the perceptual space of speed (more sources will have generated image sequences that traverse the retina more slowly than the retinal speed of the stimulus in question), the magnitude of the flash-lag effect should increase as a function of image speed.

Despite this successful prediction, there are serious concerns with this approach to understanding the speeds we see. Foremost is the adequacy of a simulated environment in determining the frequency of occurrence of different image sequences, a necessary approach because of the inability of present technology to glean this information in nature. A variety of real-world factors—gravity, the bias toward horizontal movements arising from the surface of Earth, and many others—were not included in the simulation. However, these influences are less important in the determination of image speeds than might be imagined. The overwhelming influence of perspective, which the simulation captures quite well, renders the approach relatively immune from the effects of these omissions. Because object speeds projected on a plane produce image speeds that are always less than or equal to the speeds of objects in 3-D space (see Figure 12.1), the corresponding speeds on the retina are strongly biased by perspective toward slower values. This bias is readily apparent in statistical analyses of image speeds in movies, or simply from a priori calculations. As a result, perspective is the major determinant of the frequencies of occurrence of image speeds that humans experience. Another obvious concern is how seeing motion in this counterintuitive way could possibly explain complex visually guided actions, such as hitting a fastball or returning a tennis serve. Imagining that success in the face of such challenges arises from past experience seems a stretch.

Putting these concerns aside for the moment, what about the other aspect of perceived motion—the directions of motion that we see? Can this further characteristic of motion also be explained empirically? In exploring this question, another class of motion anomalies was especially useful, namely the changes in apparent direction that occur when moving objects are seen through an occluding frame (called an "aperture" in the jargon of the field). For example, when a rod oriented at 45° moving physically from left to right at a constant speed is viewed through a circular opening that obscures its ends, its perceived direction of movement changes from horizontal to downward at about 45° from the horizontal axis (Figure 12.5). This change in direction occurs instantaneously when the frame is applied. Stranger still, the direction seen depends on the shape of the frame. For example, if the same oriented rod moving from left to right is seen through a vertically elongated rectangular frame (a vertical "slit"), the perceived direction of motion is nearly straight down. Psychologist Hans Wallach first studied these dramatic changes in perceived direction when he was a graduate student in Berlin some 70 years ago using rods moved by hand behind cardboard frames. These robust aperture effects demand some sort of explanation, and, as with the flash-leg effect, had been the subject of many studies and debates.

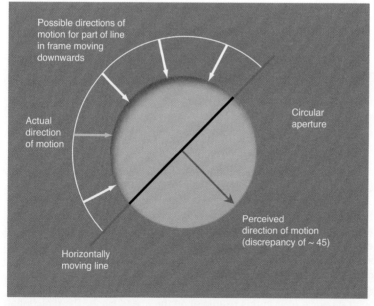

Figure 12.5   The effects of an occluding frame on the perceived direction of motion. The linear object in the aperture (the black line) is moving horizontally from left to right, as indicated by the yellow arrow. When viewed through the aperture, however, the rod or line is seen moving downward to the right (red arrow). (From Purves and Lotto, 2003)

A good place to begin thinking about the effects of apertures in empirical terms is the frequency of occurrence of the projected directions of *unoccluded* lines that objects moving in 3-D space generate (Figure 12.6A and 12.6B). Figure 12.6C shows the frequency of occurrence of fully visible lines moving in different directions on the

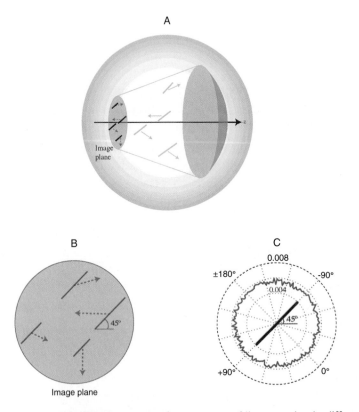

Figure 12.6   The frequency of occurrence of lines moving in different directions projected onto an image plane (a proxy for the retina) in the absence of an occluding frame. A) Diagram of a simulated environment showing rods with different orientations in space moving in different directions, but all projecting in the same orientation. B) Enlargement of the image plane in (A); notice that the projected lines move in directions that are different from the movements of the objects in space. C) The jagged gray circle indicates the distribution of the directions of movement of lines or rods such as those in (B) determined by sampling all the sequences generated by objects in the virtual environment projecting at an angle (45° in this example) on the image plane. The directions of movement are indicated around the perimeter of the graph; the distance from the center to any point on the jagged gray line indicates the frequency of occurrence of image sequences moving in that direction. (After Sung, et al., 2009. Copyright 2009 National Academy of Sciences, U.S.A.)

image plane generated by moving objects. As expected, the movement of a projected line with a given length and orientation on the image plane occurs about equally in all possible directions. This uniform distribution of image directions in the absence of occlusion describes, to a first approximation, how humans have always experienced the retinal sequences that moving rods or lines generate when they are in full view.

When a line moves behind an aperture, however, this uniform distribution of image directions changes. The different experience of projected directions that humans will always have experienced as a result of an occluding frame offers a way of examining whether the perceived directions elicited by different aperture shapes can be accounted for in wholly empirical terms. If the perceptual space for direction of motion is determined by past experience, then effects of any particular aperture should be predicted by the frequency of occurrence of the various directions generated by object motion projected through that frame.

The simplest effect to test in this way is the altered direction of motion induced by a circular aperture (Figures 12.5 and 12.7A). The frequency of occurrence of projected directions that humans will always have experienced in this circumstance can be approximated by systematically applying a circular template to the image plane of the virtual environment illustrated in Figure 12.6A and tallying the frequency of occurrence of the 2-D directions of the lines that generate image sequences within the aperture. Considering the perception of a line moving in some direction from left to right, only half the projected directions are possible (projected lines moving from right to left will never contribute to the pertinent distribution, whether looking through an aperture or not; see Figure 12.5). More important, the frequency of occurrence of lines that can move across a circular aperture with both ends occluded is strongly biased in favor of the direction orthogonal to the line (Figure 12.7B; the geometrical reasons for this are explained in Figure 12.9). Therefore the mode of this distribution (red arrows in Figure 12.7B) is the direction humans will have experienced most often whenever moving rods or lines are seen through a circular aperture. In an empirical framework, this experience should predict the direction that observers report. The green arrows in Figure 12.7B are the directions that subjects saw in psychophysical testing. As is apparent, the predicted directions (the red arrows) closely match the directions actually seen.

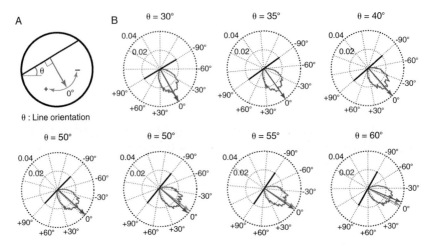

Figure 12.7   Comparison of psychophysical results and empirical predictions of the perceived directions of moving lines in different orientations seen through a circular aperture. A) As in Figure 12.6, the orientation (θ) of the line in the aperture is measured from the horizontal axis; the direction of movement is shown as a positive or negative deviation from the direction perpendicular to the moving line (0°). B) The ovoids described by the jagged gray lines are the distributions of the projected directions constrained by a circular aperture for the orientations indicated. The green arrows show the perceived directions reported in psychophysical testing, and the red arrows show the directions predicted empirically. (After Sung, et al., 2009. Copyright 2009 National Academy of Sciences, U.S.A.)

Because the perceived direction of motion of a line traversing a circular aperture can be accounted for in several other ways, the close correspondence of the observed and predicted results in Figure 12.7 is not as impressive as it seems. More convincing would be how well this (or any other) approach explains the effects on perceived direction produced by other apertures. A case in point is oriented lines or rods traveling in a horizontal direction across a vertical slit; as mentioned earlier, the aperture in Figure 12.8A generates an approximately vertical perception of movement. A more subtle effect is also apparent: As the orientation of the line becomes steeper, the perceived downward direction increasingly deviates from straight down

(Figure 12.8B). To test the merits of an empirical explanation in this case, Sung and Wojtach determined the frequency of occurrence of projected directions as a function of orientation using a vertical slit template applied millions of times to different locations across the image plane of the virtual environment. An empirical analysis again predicted the effects apparent in psychophysical testing (compare the directions of the red and green arrows in Figure 12.8). Similar success in other frames that produce peculiar effects on perceived direction (such as a triangular aperture) further buttressed the case that the directions of object motion seen are generated empirically.

The account so far begs the question of *why* the frequencies of occurrence of image sequences observed through differently shaped frames change in the ways they do. To understand the reasons for the empirically determined distributions in Figures 12.7 and 12.8, consider the biased directions of image sequences projected through a circular aperture. Figure 12.9A illustrates that for a line in any

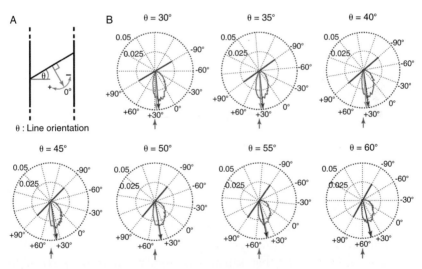

Figure 12.8   Comparison of the psychophysical results and empirical predictions of the perceived directions of moving lines in different orientations seen through a vertical slit. A) The aperture. B) Distributions of the frequency of occurrence of the projected directions of moving lines in different orientations (jagged gray ovals) when constrained by a vertical aperture. As in Figure 12.7, the green arrows are the results of psychophysical testing and the red arrows are the empirical predictions (the gray arrows indicate vertical). (After Sung, et al., 2009. Copyright 2009 National Academy of Sciences, U.S.A.)

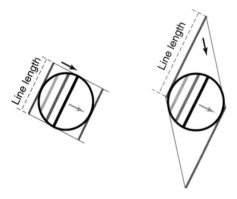

Figure 12.9 Explanation of the biased generated by the projection of lines onto an image plane such as the retina. Because a line of any given length includes all shorter lines, the occurrence of projections that fill a circular aperture generated by the red line moving in the direction indicated by the black arrow on the left will be always be more frequent than the projections generated by the blue line moving in any other direction, indicated by the black arrow on the right. As a result, the most frequently occurring projected direction when linear objects are seen through a circular aperture will always be the direction orthogonal to the orientation of the projected line (gray arrows), but not for other apertures (see Figure 12.8). (After Sung, et al., 2009. Copyright 2009 National Academy of Sciences, U.S.A.)

particular orientation on an image plane, there is a direction of motion (black arrow) that entails the minimum projected line length (the red line) that can fully occupy the aperture. A projected line traveling in any other direction (such as the blue line) must be longer if it is to fill the aperture; one end of any shorter line moving in that direction will fall inside the aperture boundary at some point, producing a different stimulus and a different perceived direction of motion. Because a line of any length includes all shorter lines, far more lines that satisfy the aperture are projected moving orthogonally to their orientation than lines moving in other directions. As a result, the distribution of directions that satisfies a circular aperture is strongly biased in the direction orthogonal to the orientation of any line, as indicated by the distributions (the jagged ovoids) in Figure 12.7B. These facts about projective geometry explain the major bias that a circular aperture produces and therefore the way humans have always experienced image sequences of linear objects moving behind a frame of this sort.

The upshot is that any aperture will produce a bias in the frequency of directions experienced that depends on the shape of the frame. But understanding the details for various apertures can be tricky. For example, perspective also requires that images of objects have dimensions that are equal to or smaller than the dimensions of physical objects that produce them. As a result, the oval-shaped distributions of the 2-D directions of lines translating in the apertures shown in Figures 12.7 and 12.8 are narrower toward the center of the graph than would be expected from the effect illustrated in Figure 12.9A alone. For a circular aperture, this additional influence does not affect the mode of the distribution, which remains orthogonal to the moving line no matter how the line is oriented on the image plane. However, for a vertical slit aperture, this addtional bias causes the frequency of occurrence of the projected directions to change as a function of the orientation of the line in the image sequence (see Figure 12.8B). The reason is that although the minimum projected line length and distance traveled needed to satisfy a circular aperture are identical, they are not the same for a vertical slit aperture. As a result, the generation of more short lines and travel distances arising from perspective for a vertical slit varies as the orientation of the line in the slit changes, explaining the empirical biases apparent when moving lines in different orientations are projected through a vertical slit.

Taken together, these empirical explanations of flash-lag and aperture effects argue that human experience with retinal projections of moving objects determines the speeds and directions we actually see. This idea is hard to swallow, and quite different from other explanations proposed for these effects and for motion perception in general. With respect to the apparent speed of moving objects, two types of theories have been offered to explain the discrepancies between physical and perceptual speeds that are apparent in the flash-lag effect and related phenomena: A misperception of objects in time versus a misperception of objects in space. The misperception in time theory proposes that the visual system compensates for neuronal processing delays by adjustments in the apparent time at which an object is seen with respect to an instantaneous marker (the flash) by anticipating or predicting the processing delays (remember that action potentials are conducted relatively slowly along axons). In contrast, the spatial theory suggests that vision employs ongoing motion information to "postdict" the position of moving objects, thereby "pushing" the apparent position of an object to a point in its trajectory that will more closely accord with its physical location when a behavioral response occurs. However, both these proposals assume that the perceived discrepancies derive from an analysis of the features of image sequences on the retina. This interpretation ignores the fact that no direct analysis of 2-D speed can specify the 3-D speed that produced it (see Figure 12.1). As a result, seeing speeds in the way these theories propose could not successfully guide behavior directed toward objects moving in the real world.

Attempts to rationalize the perceived directions of moving objects projected through apertures have taken a different approach. The most popular explanation supposes that the visual system calculates the local velocity vector field in an image sequence. The gist of this idea is that the ambiguous direction of a moving line seen through an aperture is resolved by computations based on prior knowledge about how the speeds and directions of the line in the aperture are related to the 3-D world. This approach is partly empirical, but it fails to recognize or explain the directions observers actually see in a variety of different apertures (such as the subtle effects of vertical apertures; see Figure 12.8). This theory also lacks a biological rationale.

Finally, consider again how seeing speeds and directions on a wholly empirical basis could enable us to succeed in a demanding motor task such as hitting a fastball. At first blush, it seems impossible that complex motor behaviors are made possible by trial-and-error experience instantiated in the inherited brain circuitry of the species and refined by the trial-and-error experience of individuals. This prejudice is no doubt one reason why people have regarded phenomena such as flash-lag and aperture effects as perceptual anomalies ("illusions") and not signatures of the historical way the visual system generates all motion percepts. But if one takes to heart the facts illustrated here, there is no clear alternative to this way of linking perceptions with successful actions. Although empirically determined speeds and directions can provide only an approximate guide to behavior, they will always tie necessarily uncertain image sequences to motor responses that have a good chance of succeeding. When the equally empirical but differently determined influences of other visual qualities (and the qualities associated with other sensory modalities) are taken into account, the motor responses we make have a reasonable chance of hitting a pitch or returning a serve, and an excellent chance of successfully handling more ordinary challenges. Indeed, perception of motion on an empirical basis works so well in guiding behavior it is hard to believe that the motions we see are not the motions of objects in the world.

13

How brains seem to work

To go back to the beginning, the foundation for thinking about the nervous system when I was a first-year medical student in 1960 was primarily knowledge about nerve cells and neural signaling based on work that Hodgkin, Huxley, Kuffler, Katz, and their collaborators had carried out during the preceding decades. This seminal body of research had led to a good, if incomplete, understanding of how action potentials carry information in the nervous system and how chemical synaptic transmission conveys this information from one neuron to another. Although a wealth of important detail about these processes has been added since, this basic understanding of neural function remains much the same as when I first learned it. The other body of knowledge for my generation was the welter of facts about the anatomy of the brain and the rest of the nervous system that grew concurrently. But in contrast to the clarity of what was known about neural signaling, the functional significance of brain structures and their complex interconnections was—and still is—poorly understood. The problem, both in the 1960s and today, is how brains use neural signaling to accomplish all the remarkable functions we take for granted.

It seemed obvious when I was starting out that the goal of neuroscience in the coming decades would be to put meat on the bones of the generally vague notions about what sensory, motor, and other brain systems are actually doing. As it turned out, the most influential pioneers in carrying this program forward were David Hubel and Torsten Wiesel, whose work in the visual system was inspired by Kuffler's study of the retina in the early 1950s, and by Vernon Mountcastle's description of columns in the somatic sensory cortex (see Chapter 7). By the 1970s, recording neural signals from the brains of anesthetized animals while activating a system of interest with increasingly complex stimuli had become a staple of work on sensory and motor systems. In addition to Hubel and Wiesel's work on vision, each of these efforts—mainly in the somatic sensory and auditory systems—soon had its own history and major players. Other researchers used the same general approach to explore the motor system, building on Sherrington's work earlier in the twentieth century, to ask how the motor regions of the brain are organized to produce actions. Beginning in the 1980s, understanding brain systems was further advanced by the introduction of techniques for simultaneously recording from dozens of electrodes implanted in different brain regions, and by increasingly powerful noninvasive brain imaging techniques that reveal brain activity in human subjects carrying out cognitive and other tasks. The latter approach, in particular, provided a much stronger link between the functional architecture of the brain and the inchoate anatomy that as students we were taught by rote.

Each of these ways of investigating the brain has helped close the gap between the reasonably complete understanding of neural signaling and the much less perfect grasp of how the brain uses these basic mechanisms to achieve its biological goals. But if one were to ask today what all this information indicates about how brains accomplish what they do, most neuroscientists would probably agree that the answer is not all that much clearer than it was in 1960. Why, then, despite the enormous increase in information about the properties of neurons in different regions, their connectivity, their transmitter pharmacology, and the behavioral situations in which they become active, do we remain relatively ignorant about perception, cognition, consciousness, and other brain functions that interest us most?

Part of the reason is the absence of some guiding principle or principles that would help to understand the neural underpinnings of perceptual, behavioral, and cognitive phenomenology in a more general way. One doesn't have to be steeped in history to recognize that other domains of biological science have typically advanced under the banner of some overarching framework. Obvious examples are the cell theory as a principle for understanding the structure and function of organs that emerged in the seventeenth century, the theory of evolution by natural selection as a principle for understanding species diversity that emerged in the nineteenth century, and the theory of genes as a principle for understanding inherited traits that emerged in the late nineteenth and early twentieth centuries. However, in the case of the brain, no overarching idea has arisen that serves in this way. Given the complexity of the brain and the rest of the nervous system, this absence is perhaps to be expected. But if the history of science is any guide, sooner or later a principle of this sort is bound to emerge.

Only a few ideas about the overall operation of the brain have been seriously considered in recent decades, and all have relied heavily on studies of vision and the visual brain. The one that has come up most often in earlier chapters is that the brain extracts features from sensory stimuli and combines them in representations of the world in higher-order cortical areas that we then perceive and use to inform behavior and cognition. This perspective takes the brain to be directly related to the real world by logical operations carried out by neurons, neural circuits, and brain systems. The explicit or implicit advocacy of this essentially rationalist interpretation of brains has focused on both the biology of neural processing and the logical instructions (computational algorithms) that might be doing the job. A more stochastic twist on this perspective is that because stimuli are inevitably noisy and their sources are uncertain, the brain must to a degree operate probabilistically, generating perceptions and behaviors predicated on inferences (best guesses) about the most likely state of the world. These frameworks for thinking about how brains work express in different ways the seemingly obvious idea that sensory information is processed by online neural computations that represent the "real world" somewhere in the brain as a basis for all the higher-order functions that the brain performs and the behaviors that follow.

I fully accepted this general concept of brain function gleaned from my professors and colleagues in the 1960s. This framework taps into the subjective sense we all have about our relationship to the world, and for the next 20 years or so, I had no particular reason to doubt its propriety. Indeed, it seemed clear that exploring the properties of the nerve cells and neural circuits from the bottom up would eventually reveal the underlying logic of vision and the other senses, and indicate how they were related to perception and, ultimately, action. This was the implicit message of Hubel and Wiesel's effort to understand vision in terms of an anatomical and functional hierarchy in which simple cells feed onto complex cells, complex cells feed onto hypercomplex cells, and so on up to the higher reaches of the extrastriate cortex. (As noted earlier, Hubel and Wiesel were wary of anything that smacked of theoretical speculation and wisely stuck close to their results, leaving others to argue about what they might mean for the function of the brain more generally.) Nearly everyone believed that the activity of neurons with specific receptive field properties would, at some level of the visual system, represent the combined image features of a stimulus, thereby accounting for what we see. This idea seems so straightforward that it was, and is, hard to imagine an alternative.

Nevertheless, this conception of how brains work has not been substantiated despite an effort that now spans 50 years. When a path in science is pursued for this long without the emergence of a deeper understanding of the issue being addressed, doubts are usually warranted. Since the 1980s, it has become increasingly apparent that neuronal responses to even the simplest visual stimuli are difficult to rationalize in terms of a hierarchy that begins with the detection of image features at the retinal level and ends with feature representation in the cerebral cortex. Another problem has been the growing realization that the activity of visual neurons is not just a result of whatever is going on within the small region of visual space that defines the "classical" receptive field of a visual neuron; stimuli that fall outside this region also affect the activity of neurons in ways that have been difficult to understand. This less direct source of activation and its complexity is not surprising, given that the vast majority of the inputs to neurons in the various stations of the visual system come not from the retina, but from other neurons at the same or higher levels (such as feedback to the thalamus from the primary visual cortex, and feedback to the primary visual cortex from extrastriate cortical regions). Yet another concern is the observation that different stimuli can elicit the same general pattern of visual cortical activity. If the evidence in previous chapters that the stimuli can be perceived in radically different ways as a function of context is added to this mix, one is left with deep uncertainty about how the activity of sensory neurons produces perceptions, and how perception is then related to action.

One way to deal with these concerns was to take a frankly computational approach to the brain, a view that was gaining popularity during the same time that visual physiologists were grappling with these problems. The idea was that if brains operate as computers do, then understanding brain functions might advance rapidly by riding on the back of computational theory. The potential of computers and computational theory had gotten off to a rough start with the grandiose and ultimately failed plans of Charles Babbage in the nineteenth century (Figure 13.1), but the enormous progress of computation during and after World War II encouraged its application to understanding brain functions, vision in particular. The most influential exponent of visual processing as computations that could be understood in logical terms was David Marr. A brilliant theorist, Marr summarized his ideas in a book published in 1982 that many regard as having given birth to the field of computer vision. Computer (or "machine") vision is the attractive notion that one should be able to build and program a device that sees in much the same way we do, based on rational engineering principles. For Marr (and many others since), vision is a computational task that, similar to the framework underlying the physiological approach that was then in ascendancy, leads to a representation of stimulus features in the brain. Marr (who died of cancer at age 35 just before his book was published) integrated into this perspective much of what was then known about visual perception, physiology, and anatomy. He was motivated by his conviction that vision scientists were describing the physiological and anatomical underpinnings of vision but were missing the boat by not getting straight to the logical heart of the matter.

To address this deficiency, Marr proposed that the information in retinal images is subject to algorithmic processing at three levels, which he referred to as the construction of the primal sketch, the 2½-D sketch, and the 3-D model representation. The purpose of the primal sketch is to make information about the elemental geometries and intensities of the retinal image explicit (to represent basic image features), the purpose of the 2½-D sketch is to make explicit the orientation and approximate depth of surfaces in a viewer-centered frame of reference, and the purpose of the 3-D model is to represent shapes and their spatial organization in an object-centered frame of reference, representing what observers see. For Marr, this overall process of internal representation was a "formal system for making explicit certain entities or types of information, together with a specification of how the system does this." His attempt to rationalize vision in a

Figure 13.1 Babbage's "Difference Engine," a mechanical computer designed in 1821. Although this and Babbage's later designs were never built to completion during his lifetime (his machinist eventually quit and the British government tired of supporting his increasingly expensive project), Babbage's machines are widely considered the first examples of serial-processing computers, the computers on which most contemporary machine-vision strategies are implemented. Because its mechanisms are visible, Babbage's machines (some of which are on display at the British Science Museum in London) give a more vivid impression of what logical computation entails than looking at an integrated circuit. (Courtesy of Eric Foxley)

comprehensive computational theory was highly ambitious and widely admired. Although much of this initial enthusiasm has faded, determining how the features of the retinal image are detected, processed, and represented according to a series of algorithms remains a central theme in vision science, particularly in the ongoing attempts of computer scientists, biomedical engineers, and the military to create artificial visual systems that could help patients or enable robotic devices to navigate more efficiently in real-world environments.

Wherever Marr might have taken these ideas had he lived, it seems unlikely that the brain operates like the serial-processing computers we are all familiar with today—a machine that uses a series of specified, logical steps to solve a problem (a so-called universal Turing machine). Of course, algorithmic computations based on physical information acquired by photometers, laser range scanners, or other devices that directly measure aspects of the world can solve some of the problems that confront biological vision. And the "visually guided" behavior of robotic vehicles today is impressive. But these and other automata are "seeing" in a fundamentally different way than we do. The limitation of machine vision in this form is its inability to meet the challenge that has evidently driven the evolution of biological vision: The unknowability of the physical world by any direct operation on images (the inverse problem). Machines such as photometers and laser range finders accurately determine some physical property of the world (such as luminance or distance) by direct measurement. But as should be apparent from previous chapters, this is not an option for biological vision, nor for machine vision if it is ever to attain the sort of visual competence that we enjoy. Only by evolving circuitry that reflects the outcome of trial-and-error experience with all the variables that affect the successful behavior is a machine likely to generate "perceptions" and "visually guided" behavior that works well in real-world circumstances.

Given this caveat, it is important to recognize that computers can solve complex problems in another way, an alternative that gives cause for some optimism about the future of machine vision and ultimately understanding what the complex connectivity of the brain is accomplishing. In a paper published in 1943, MIT psychologist Warren McCulloch and logician Walter Pitts pointed out that instead of depending on a series of predetermined steps that dictate each sequential operation of a computer in logical terms, problems can also be solved by devices that comprise a network of units (*neurons*, in their biologically inspired terminology) whose interconnections change progressively according to feedback arising from the success (or failure) of the network dealing with the problem at hand (Figure 13.2). The key attribute of such systems—which quickly came to be called artificial neural networks (or just neural nets)—is the ability to solve a problem without previous knowledge of the answer, the steps needed to reach it, or the designer's conception of how the problem might be solved in rational terms. In effect, neural nets reach solutions by trial and error, gradually generating more useful responses by retaining the connectivity that led to improved behavior. As a result, the architecture of the trained network—analogous to evolved brain circuitry—is entirely a result of the network's experience.

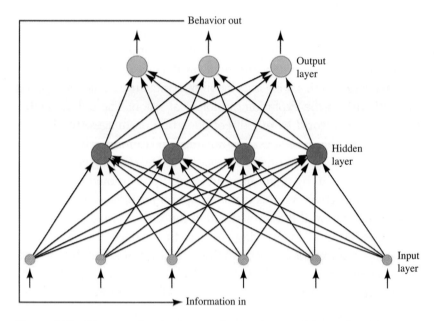

Figure 13.2   Diagram of a simple neural network comprising an input layer, an output layer, and a hidden layer. The common denominator of this and more complex artificial neural networks is a richly interconnected system of nodes, or neurons. The strengths of the initially random connections between the nodes are progressively changed according to the relative success of trial-and-error responses, which is then fed back to the network during training. The result is that the connectivity of the system is gradually changed as the network deals ever more effectively whatever problem it has been given. (After Purves, Brannon, et al., 2008)

The answer to the question of whether the brain operates like a computer therefore depends on what kind of a computer one has in mind. Unlike the rationalist idea of computation based on feature extraction and representation according to a series of logical steps, the conception of the brain as a computer is plausible considered as a neural network whose connectivity changes over evolutionary and individual time according to feedback from trial and error. This perspective can be made even more biologically attractive by letting a computational equivalent of natural selection operate on an evolving population of neural networks. By changing the weights of the connections between nodes according to the relative success of autonomous networks competing with each other in a virtual environment and reproducing differentially based on some criteria of "fitness," it should be possible to explore how the structure and function of real brains evolved.

By the mid-1990s, mathematically inclined psychologists joined the fray with yet another perspective about how vision, and brains more generally, might work. Their approach was based on a statistical methodology generally referred to as Bayesian decision theory. Thomas Bayes was an English mathematician and Presbyterian minister who published a paper in 1763 showing formally how conditional probabilities lead to valid inferences. Although Bayes's motivation for studying this issue is unclear, his theorem has been widely applied to statistical problems whose solution depends on an assessment of hypotheses that are only more or less likely to be true because they depend on two or more probabilities. For example, the theorem has been used in medicine to evaluate the likelihood of a diagnosis or a clinical outcome given a set of tests or other data that together contribute to the overall probability of having the disease or responding to therapy. Bayes's theorem is usually written in this form:

$$P(H|E) = \frac{P(H) * P(E|H)}{P(E)}$$

H is a hypothesis to be tested, E is the evidence pertinent to its validity, and P is the probability. The first term on the right side of Bayes's equation, P(H), is called the prior probability distribution, or simply "the prior," and is a statistical measure of confidence in the hypothesis, without any present evidence about its truth or falsity. With respect to vision, the prior describes the probabilities of different physical states of the world that might have produced a retinal image. Therefore, the needed priors are the frequency of occurrence in the world of surface reflectance values, illuminants, distances, object sizes, and so on. The second term, P(E|H), is called the likelihood function. If hypothesis H was true, this term indicates the probability that the evidence E would have been available to support it. Given a particular state of the physical world (a particular combination of illumination, surface properties, object sizes, etc.), the likelihood function describes the probability of that state having generated the retinal projection in question. The product of the prior and the likelihood function, divided by the normalization constant P(E), gives the posterior probability distribution, P(H|E). The posterior distribution defines the probability of hypothesis H being true, given the evidence E, and would therefore indicate the relative probability of a given retinal image having been generated by some set of possible physical realities. The most likely physical reality in the posterior probability distribution is taken to be what an observer sees.

Despite the difficulty of these more abstract ideas, Bayes's theorem correctly (and usefully) spells out the logical relationship among the factors that underlie any empirical approach to vision. Nonetheless, the way it has been used in vision science has tended to obscure instead of clarify the way brains seem to work. For most people using Bayesian decision theory, the visual system is taken to compute statistical inferences about possible physical states of the world based on prior knowledge acquired through experience with image features. The evidence in the preceding chapters, however, argues that the accumulated knowledge stored in brain circuitry is based on feedback from behavioral success, not image features. As a result, what we see is not the state of the world that most likely produced the image. On the contrary, the qualities we see define a subjective universe that comprises the various perceptual spaces described earlier. These perceptions successfully guide behavior not because they represent likely states of the world but because of the trial-and-error manner in which the relevant perceptual spaces and the underlying neural circuitry have been generated.

Although the human brain is enormously complex, the evidence based on what we see and why implies that the brain and the rest of our nervous systems are doing just one basic thing: linking sensory information (perceptions or the unconscious equivalent) to successful behavior by means of synaptic connectivity that has been entirely determined by trial and error. The reason for this blunt assertion is that the inverse problem in vision and other sensory modalities doesn't allow much choice.

The biological mechanisms that create neural circuitry on this basis are straightforward, at least in principle. Whenever the neural connections inherited by an individual link perception and action a little more successfully, the slightly greater reproductive fitness of the individual tends to increase the prevalence of that brain circuitry in the population. Over eons, the neuronal associations underlying successful behavior in response to stimuli will therefore increase and unsuccessful associations will decrease, leading to the brain circuitry humans and other animals have today. All that is needed to implement this strategy is plenty of time and a way of providing feedback about the relative success of behavior. The history of life on Earth has provided ample time (approximately 3.5 billion years), and natural selection is a wonderfully powerful mechanism for assigning credit to behavior. Indeed, our perceptions and actions have become so effective in dealing with our circumstances world that it is difficult to imagine that what we see, hear, or otherwise perceive is only an operationally useful surrogate for the world, not physical reality.

The critical importance of this inherited neural infrastructure for survival is self-evident. Nonetheless, the neural connectivity we are born with is continually refined by individual experience (see Chapters 2–7). The reasons for this are also clear. From a developmental perspective, neural circuits must adjust to the changing body, as well as reaping the obvious benefits that come from encoding additional information about the specific objects, conditions, and contingencies encountered in any particular life. Although the mechanisms of synaptic plasticity that enable lifetime learning are quite different from the mechanisms of inheritance, both serve to encode experience in neural circuitry. The common denominator is the association of input from the external and internal environment with empirically successful behavioral output. From this perspective, the complexity of brain function and structure boils down to using, making, and modifying neuronal connections. The best evidence so far that brains really do work in this wholly empirical way is that so many peculiarities of subjective experience can be explained in this way. For each of the basic visual qualities—lightness, brightness, color, geometrical form, and motion—a variety of otherwise puzzling phenomena can be predicted by the frequency of occurrence of the stimulus characteristics derived from databases that serve as proxies for human experience. The universal discrepancies between perception and physical measurement (the more flagrant examples of which are wrongly categorized as "illusions") are neither anomalies nor the result of the limitations in neural processing circuitry. They are the signatures of this strategy of brain operation. Whereas the end result gives us a compelling sense that our brains must be detecting features by analyzing stimuli and producing perceptions that represent the world, terms such as feature detection, feature analysis, and feature representation are not appropriate descriptors of what the machinery of the brain is doing, the problem this mode of operation is solving, or what we actually perceive.

Accounting in these terms for everything we see or otherwise perceive, however, is going to be much harder than suggested by the ability to predict the relatively simple visual responses used so far to support the empirical case. Even in the most basic circumstances, what we see with respect to any particular visual quality is, for good reason, affected by the sensory information pertinent to other qualities (see Chapter 9). To complicate matters further, such interactions extend across different sensory modalities for the same reasons. One of many examples of how what we hear affects what we see is illustrated in Figure 13.3. This further evidence implies that percepts and behaviors ultimately depend on neural processing that is going on at the same time in many brain systems, taking into account the influence of all the neural activity elicited by the stimuli acting on us at any given time. The advantages accruing from such interactions help explain why the connectivity of the brain is so complex. Any brain that did not take into account the full range of information available to it to determine behavior would not be doing its job. If one adds the influences of memory, motivation, and emotional state to sensory input, what we perceive and do at any given moment is determined by the processing going on in much, if not all, of the brain.

This understanding of perception and sensory systems generally implies yet another difficult conclusion about how brains work: Perceptions, and the behaviors they lead to, are entirely reflexive. Although the concept of a reflex is not precisely defined in neuroscience, the term is typically used to refer to involuntary sensory-motor behaviors that are assumed to occur with a minimum of the higher-order cortical processing thought to underlie voluntary

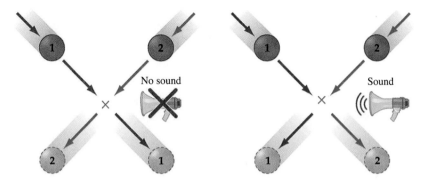

Figure 13.3   Example of how what we hear influences what we see. The perceived trajectories (colored arrows) of two objects (1, 2) can be altered by the sound of a collision coincident with the moment the objects would have bumped into each other. (After Purves, Brannon, et al., 2008)

actions. The usual example presented in textbooks is one of the spinal reflexes that Sherrington studied in the early twentieth century (Figure 13.4), or an autonomic reflex such as salivation in response to food that Ivan Pavlov used at about the same time to study conditioned learning. For Sherrington and Pavlov, a reflex meant an automatic response that depends on a relatively simple neural pathway with little or no cognitive input. However, Sherrington was well aware that the idea of a simple reflex isolated from the rest of the nervous system is, in his words, a "convenient...fiction." Sherrington went on to emphasize that "all parts of the nervous system are connected together and no part of it is ever capable of reaction without affecting and being affected by other parts". Clinicians know very well that spinal reflexes depend on cortical processing and that they no longer operate normally when the descending cortical input to spinal neurons is compromised by disease or injury.

Despite Sherrington's caveat, reflexes are usually regarded as behaviors that are immune from the influence of cortical processing, particularly from "top-down" cognitive influences. But the reality is that brain systems and circuits are invariably interconnected to take full advantage of the range of information the nervous system provides at any given moment. From this vantage, the conventional distinction between involuntary (reflexive) and voluntary neural processing makes little sense. Any perception or behavior is equally a product of *all* the neural processing occurring in the nervous system

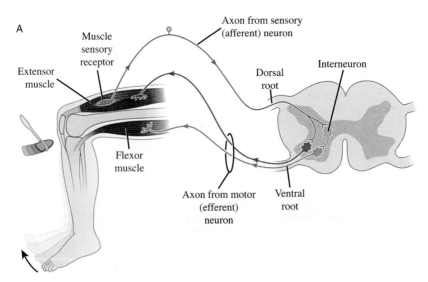

Figure 13.4   Diagram of a "simple" spinal reflex. The pathways that carry information from the spinal cord to the brainstem and cerebrum, and the descending pathways from the brain that modulate the execution of this or any other behavior, are not shown in this diagram of the knee-jerk reflex. (After Purves, Brannon et al., 2008)

at that time. Depending on the circumstances, some parts of the nervous system will be more—or less—active than others. But it would be a mistake to think that processing activity in any part of the brain is irrelevant to processing in another region, or to perceptual and behavioral outcome. A subjective sense of what our brains are doing has given pride of place to what we assume to be "voluntary" behaviors or thought processes, but no neurobiological evidence supports this bias.

The reflexive nature of perception and behavior points to a still deeper problem at the center of a debate that has persisted since the Stoic and Epicurean philosophers were at each other's throats over this issue in ancient Greece: Does the way brains work allow us to act freely, or is what we think and do fully determined? The evidence discussed here and in previous chapters inevitably bears on this vexing question. Based on empirically determined sensory circuitry, the brain generates behavior that works surprisingly well in a world that our senses can't tell us about directly. This way of producing perceptual content and behavior is very different from our subjective impression of what is going on. Everyday experience suggests that our brains (a stand-in for the "me" or the "I" that we take for granted) analyze stimulus features and represent these features, and that we then make decisions about how to act based on representations of the world "as it really is." But if brain activity and its perceptual and behavioral consequences simply reflect neural associations created by accumulated phylogenetic and ontogenetic experience, then a wholly empirical understanding of brain function comes down squarely on the side of determinism. In this framework, terms such as *inferences, decisions,* and *choices* are apt descriptors of our subjective sense of the way we relate to the world, but not of any underlying brain function.

* * *

To sum up how brains seem to work, the circuitry of nervous systems such as ours has evolved to contend with one fundamental challenge: How to generate useful perceptions and behaviors in response to a world that is unknowable directly by means of sensory stimuli. The strategy that has emerged to deal with this problem is governed by history, not logical principles or algorithms. Based on feedback from the empirical consequences of behavior, accumulated information about operational success is realized over evolutionary time in inherited neural circuitry whose organization is then modified to a limited extent by individual experience. Accordingly, our perceptions never correspond to physical reality despite the fact that they provide successful operational guides to behavior. The evidence that supports these conclusions is the ability to predict many otherwise puzzling perceptual phenomena using databases that serve as proxies for aspects of accumulated human experience.

These ideas about what we perceive, what we do as a result, and ultimately what we are in consequence may be anathema to neuroscientists committed to a rationalist perspective of brain function, or to anyone who has difficulty disengaging from the subjective sense we have of the world and our relation to it. But if the evidence continues to support a wholly empirical interpretation of how brains work, we will simply need to pursue brain function and structure in these terms, perhaps gaining in the process a clearer and more useful conception of ourselves and our place in nature.

Suggested reading

Adelson, E. H. "Light Perception and Lightness Illusions." In *The New Cognitive Neurosciences*, 2nd edition, edited by M. S. Gazzaniga. Cambridge, MA: MIT, 2000. 339–351.

Allman, J. M. *Evolving Brains*. New York: W.H. Freeman, 1999.

Barlow, H. B. "The Neuron Doctrine in Perception." In *Cognitive Neurosciences*, edited by M. S. Gazzaniga. Cambridge, MA: MIT Press, 1995. 415–436.

Berkeley, G. *Philosophical Works Including Works on Vision*, edited by M. R. Ayers. London: Everyman/ J. M. Dent, 1975.

Blake, R. "A primer on binocular rivalry, including current controversies." *Brain and Mind* 2 (2000): 5–38.

Bowmaker, J. K. "The Evolution of Color Vision in Vertebrates." *Eye* 12 (1998): 541–547.

Chevreul, M. E. *The Principles of Harmony and Contrast of Colours, and Their Applications to the Arts*, 2nd ed. Translated by C. Martel. London: Longman, 1987.

Cornsweet, T. N. *Visual Perception*. New York: Academic Press, 1970.

Daw, N. W. "The Psychology and Physiology of Color Vision." Trends in Cognitive Sciences 9 (1984): 330–335.

Descartes, R. *Discourse on Method, Optics, Geometry, and Meteorology*. Translated by P. Olscamp. Indianapolis: Bobbs-Merrill, 1965.

Douglas, R. J., and K. C. Martin. "Mapping the Matrix: The Ways of the Neocortex." *Neuron* 56 (2007): 226–238.

Eagleman, D. M., and T. J. Sejnowski. "Untangling Spatial from Temporal Illusions." *Trends in Neurosciences* 25 (2002): 293.

Evans, R. M. *An Introduction to Color*. New York: John Wiley, 1948.

Fitzpatrick, D. "Seeing Beyond the Receptive Field in Primary Visual Cortex." *Current Opinion in Neurobiology* 10 (2000): 438–443.

Geisler, W. S., and D. Kersten. "Illusions, Perception, and Bayes." *Nature Neuroscience* 5, no. 6 (2002): 508–510.

Gibson, J. J. *The Senses Considered As Perceptual Systems*. Boston: Houghton Mifflin, 1966.

Gilchrist, A. *Seeing in Black and White*. New York: Oxford University Press, 2006.

Gregory, R. L. *Eye and Brain: The Psychology of Seeing*, 5th ed. Oxford: Oxford University Press, 1998.

Hering, E. "Outlines of a Theory of the Light Sense." In *Grundzüge der Lehre vom Lichtsinn*. Translated by L. M. Hurvich and D. Jameson. Cambridge, MA: Harvard University Press, 1964.

Hildreth, E. *The Measurement of Visual Motion*. Cambridge, MA: MIT Press, 1984.

Howe, C. Q., and D. Purves. *Perceiving Geometry: Geometrical Illusions Explained by Natural Scene Statistics*. New York: Springer Press, 2005.

Hubel, D. H. "Exploration of the Primary Visual Cortex, 1955–1978." *Nature* 299 (1982): 515–524.

Hubel, D. H. *Eye, Brain, and Vision*. New York: W.H. Freeman, 1988.

Hubel, D. H., and T. N. Wiesel. *Brain and Visual Perception*. New York: Oxford University Press, 2006.

Hubel, D. H., and T. N. Wiesel. "Ferrier Lecture: Functional Architecture of Macaque Monkey Visual Cortex." *Philosophical Transactions of the Royal Society B* 198 (1977): 1–59.

Hurvich, L. *Color Vision*. Sunderland, MA: Sinauer Associates, 1981.

Jameson, D., and L. Hurvich. "Essay Concerning Color Constancy." *Annual Review of Psychology* 40 (1989): 1–22.

Kersten, D., and A. Yuille. "Bayesian Models of Object Perception." *Current Opinion in Neurobiology* 13, no. 2 (2003): 150–158.

Koffka, K. *Principals of Gestalt Psychology*. New York: Harcourt Brace, 1935.

Kuehni, R. G., and A. Schwarz. *Color Ordered: A Survey of Color Order Systems from Antiquity to the Present*. Oxford: Oxford University Press, 2008.

Kuffler, S. W. "The Single-Cell Approach in the Visual System and the Study of Receptive Fields." *Investigative Ophthalmology* 12 (1973): 794–813.

Land, E. H. "Recent Advances in Retinex Theory." *Vision Research* 26 (1986): 7–21.

Luckiesh, M. *Visual Illusions: Their Causes, Characteristics, and Applications*. New York: Van Nostrand, 1922.

Mach, E. *The Analysis of Sensations and the Relation of the Physical to the Psychical*, 1st German ed. Translated by C. M. Williams. New York: Dover, 1959.

Marr, D. *Vision: A Computational Investigation into Human Representation and Processing of Visual Information*. San Francisco: W.H. Freeman, 1982.

Maxwell, J. C. "Experiments on Colour, As Perceived by the Eye, with Remarks on Colour-blindness." *Transactions of the Royal Society of Edinburgh XXI* 21 (1855): 275–298.

McCulloch, W. S., and W. Pitts. "A Logical Calculus of the Ideas Immanent in Nervous Activity." *Bulletin of Mathematical Biophysics* 5 (1943): 115–133.

Minnaert, M. G. J. *Light and Color in the Outdoors*. New York: Springer, 1992.

Neumeyer, C. "Evolution of Colour Vision." In *Uses and Evolutionary Origins of Primate Color Vision: Evolution of the Eye and Visual System*. Edited by J. R. Cronly-Dillon and R. L. Gregory. Boca Raton, FL: CRC Press, 1991. 284–305.

Olshausen, B. A., and D. J. Field. "Emergence of Simple-Cell Receptive Field Properties By Learning a Sparse Code for Natural Images." *Nature* 381 (1996): 607–609.

Palmer, S. *Vision Science: From Photons to Phenomenology*. Cambridge, MA: MIT Press, 1999.

Pavlov, I. P. *Conditioned Reflexes: An Investigation of the Physiological Activity of the Cerebral Cortex*. Translated and edited by G. V. Anrep. New York: Dover, 1960.

Purves, D., and J. Lichtman. *Principles of Neural Development*. Sunderland, MA: Sinauer Associates, 1985.

Purves, D., and B. Lotto. *Why We See What We Do: An Empirical Theory of Vision*. Sunderland, MA: Sinauer Associates, 2003.

Purves, D., G. A. Augustine, D. Fitzpatrick, W. Hall, A. S. LaMantia, J. O. McNamara, and S. M. Williams. *Neuroscience*, 4th ed. Sunderland, MA: Sinauer Associates, 2008.

Purves, D., E. M. Brannon, R. Cabeza, S. A. Huettel, K. S. LaBar, M. L. Platt, and M. G. Woldorff. *Principles of Cognitive Neuroscience*. Sunderland, MA: Sinauer Associates, 2008.

Purves, D., M. S. Williams, S. Nundy, and R. B. Lotto. "Perceiving the Intensity of Light." *Psychological Review* 111 (2004): 142–158.

Rao, R. P. N., B. A. Olhausen, and M. S. Lewkicki, eds. *Probabilistic Models of the Brain: Perception and Neural Function*. Boston: MIT Press, 2002.

Robinson, J. O. *The Psychology of Visual Illusion*. Mineola, NY: Dover, 1998.

Rock, I. *Perception*. New York: Scientific American Library, 1995.

Rodieck, R. W. *First Steps in Seeing*. Sunderland, MA: Sinauer Associates, 1998.

Shepard, R. N. *L'oeil Qui Pense: Visions, Illusions, Perceptions*. Paris: Le Seuil, coll. Points Sciences (1992): S140.

Sherrington, S. C. *The Integrative Action of the Nervous System*, 2nd ed. New Haven, CT: Yale University Press, 1947.

Sherrington, C. S. "On Reciprocal Action in the Retina As Studied By Means of Some Rotating Discs." In *Reciprocal Action in the Retina*. Cambridge UK: Cambridge University Press, 1906. 33–54.

Simoncelli, E. P., and B. A. Olshausen. "Natural Image Statistics and Neural Representation." *Annual Review of Neuroscience* 24 (2001): 1193–1216.

Stevens, S. S. *Psychophysics*. New York: John Wiley, 1975.

Sung, K., W. T. Wojtach, and D. Purves. "An Empirical Explanation of Aperture Effects." Proceedings of the National Academy of Sciences 106 (2009): 298–303.

Turing, A. "Computing Machinery and Intelligence." *Mind* 59 (1950): 433–460.

von Helmholtz, H. L. F. *Helmholtz's Treatise on Physiological Optics*. 3rd German ed. Translated by J. P. C. Southall. Rochester, NY: The Optical Society of America, 1925.

Walls, G. L. "The G. Palmer Story." *Journal of the History of Medicine and Allied Sciences* 11: 66–96.

Wertheimer, M. "Laws of Organization in Perceptual Forms." In *A Sourcebook of Gestalt Psychology*. Translated and edited by W. D. Ellis. New York: Humanities Press, 1950. 71–88.

Wheatstone, C. "Contributions to the Physiology of Vision. I. On Some Remarkable and Hitherto Unobserved Phenomena of Binocular Vision." *Philosophical Transactions of the Royal Society B* 128 (1838): 371–394.

Wiesel, T. N. "Postnatal Development of the Visual Cortex and the Influence of Environment." *Nature* 299 (1982): 583–591.

Wojtach, W. T., K. Sung, S. Truong, and D. Purves. "An Empirical Explanation of the Flash-Lag Effect." *Proceedings of the National Academy of Sciences* 105 (2008): 16338–16343.

Wuerger, S., R. Shapley, and N. Rubin. "'On the Visually Perceived Direction of Motion' by Hans Wallach 60 Years Later." *Perception* 25 (1996): 1317–1367.

Wundt, W. "Contributions to the Theory of Sensory Perception." In *Classics in Psychology*. Edited by T. Shipley. New York: Philosophical Library, 1961. 51–78.

Wyszecki, G., and W. S. Stiles. *Color Science: Concepts and Methods, Quantitative Data and Formulae*. Edited by G. Wyszecki and W. S. Stiles. New York: John Wiley, 1982.

Yang, Z., and D. Purves. "The Statistical Structure of Natural Light Patterns Determines Perceived Light Intensity." *Proceedings of the National Academy of Sciences* 101 (2004): 8745–8750.

Young, T. "An Account of Some Cases of the Production of Colours, Not Hitherto Described." *Philosophical Transactions of the Royal Society B* 92 (1802): 387–397.

Glossary

absorption spectrum

The spectral distribution of light that has passed through a medium (or been reflected from a surface) that absorbs some portion of the incident light.

achromatic

Pertaining to visual stimuli (or scenes) that are perceived only in shades of grays ranging from black to white.

action potential

The electrical signal conducted along neuronal axons by which information is conveyed from one place to another in the nervous system.

acuity

The ability of the visual system to accurately discriminate spatial detail, as in the standard Snellen eye chart exam. Usually tested by the spatial discrimination of two points.

adaptation

Adjustment to different levels of stimulus intensity, which allows operation of a sensory system over a wide range.

afferent neuron

An axon that conducts action potentials from the periphery to more central parts of the nervous system.

algorithm

A set of rules or procedures set down in logical notation, typically (but not necessarily) carried out by a computer.

amblyopia

Diminished visual acuity arising from a failure to establish appropriate visual cortical connections in early life, typically as a result of visual deprivation.

analytical (analyze)

To determine the nature of something according to a set of principles, such as the features of an image (contrasts with *empirical*).

aperture

An opening in the foreground through which one can view a scene (think of looking through a keyhole).

aperture problem

The challenge of explaining the different speed and direction of a moving line perceived through an aperture.

artificial intelligence

A phrase used to describe a computational approach to mimicking brain function that generally depends on algorithmic solutions.

artificial neural network

A computer architecture for solving problems by feedback from trial and error instead of by a predetermined algorithm.

association cortex

Regions of the cerebral neocortex defined by their lack of involvement in primary sensory or motor processing.

autonomic ganglia

Collections of autonomic motor neurons outside the central nervous system that innervate visceral smooth muscles, cardiac muscle, and glands.

autonomic nervous system

All the neural apparatus that controls visceral behavior. Includes the sympathetic, parasympathetic, and enteric systems.

awareness

Synonym for *consciousness*.

background

Referring to the part or parts of a scene that are farther away from an observer and/or less salient.

Bayesian decision theory

The application of Bayes's theorem to real-world problems.

Bayes's theorem

A theorem that formally describes how valid inferences can be drawn from conditional probabilities.

binding problem

Understanding how perceptual qualities are brought together in the perception of objects.

binocular

Pertaining to both eyes.

binocular rivalry

See *rivalry*.

blind spot

The region of visual space that falls on the optic disk in the view generated by each eye. Because of the lack of photoreceptors in this part of the retina, objects that lie completely within the blind spot are not perceived in monocular view.

blobs

Iterated units of unknown function in the primary visual cortex of humans and most other primates.

bottom-up

A term that loosely refers to the flow of information from sensory receptors to the cerebral cortex.

bottom-up processing

Generally refers to peripheral sensory processing.

brain

The cerebral hemispheres, cerebellum, and brainstem.

brainstem

The portion of the brain that lies between the diencephalon and the spinal cord. Comprises the midbrain, pons, and medulla.

brightness

Technically, the apparent intensity of a source of light; more generally, a sense of the effective overall intensity of a light stimulus. See *lightness*.

Brodmann's Area 17

The primary visual cortex, also called the striate cortex.

calcarine sulcus

The major sulcus on the medial aspect of the human occipital lobe. The primary visual cortex lies largely within this sulcus.

cell

The basic biological unit in plants and animals, defined by a cell membrane that encloses cytoplasm and (typically) the cell nucleus.

cell body

The portion of a neuron that houses the nucleus.

central nervous system

The brain and spinal cord of vertebrates (by analogy, the central nerve cord and ganglia of invertebrates).

central sulcus

A major sulcus on the upper and lateral aspect of the hemispheres that forms the boundary between the frontal and parietal lobes. The anterior bank of the sulcus contains the primary motor cortex; the posterior bank contains the primary somatic sensory cortex.

cerebellum

Prominent hindbrain structure concerned primarily with motor coordination, posture, and balance.

cerebral cortex

The superficial gray matter of the cerebral hemispheres.

cerebrum

The largest and most rostral part of the brain in humans and other mammals, consisting of the two cerebral hemispheres.

channels

Pores in the membrane of neurons and other cells that allow the inward and outward movements of ions that underlie neural signaling.

chiasm (optic chiasm)

The crossing of optic nerve axons from the nasal portions of the retinas in humans and other mammals.

circuitry

In neurobiology, refers to the connections between neurons. Usually pertinent to some particular function (as in visual circuitry).

cognition

Referring to higher-order mental processes such as perception, attention, and memory.

color

The subjective sensations elicited in humans by different distributions of power in the spectra of light stimuli.

color blind

Outmoded term for individuals who have abnormal or absent color vision. See *color deficient*.

color constancy

The similar appearance of surfaces despite different light spectra coming from them. Usually applied to the approximate maintenance of object appearances in different illuminants.

color contrast

The different appearance of surfaces despite similar spectra coming from them.

color deficient

Term for individuals who have abnormal color vision as a result of the absence of (or abnormalities in) one or more of the three human cone types.

color opponent

Term used to refer to the experience of seeing red–green, blue–yellow, and black–white as opposites.

color-opponent cells

Cells whose receptive field centers and surrounds are sensitive to opposing spectral qualities.

color space

The depiction of a human color experience in diagrammatic form by a space with three axes representing the perceptual attributes of hue, saturation, and brightness.

colorimetry

Measurements of the human responses to uniform spectral stimuli presented in the laboratory with a minimum of contextual information.

competition

In biology, the struggle for limited resources essential to survival or growth.

complementary colors

The colors elicited by lights that, when mixed together, generate a neutral sensation of some shade of gray (often applied more loosely to colors that are oppositely disposed around the Newton color circle).

cone opsins

The distinct photopigment proteins found in human and other cones.

cones

Photoreceptors specialized for high visual acuity and the perception of color.

consciousness

A contentious concept that includes the ideas of wakefulness, awareness of the world, and awareness of the self as an actor in the world.

context

The information provided by the surroundings of a target. The division of a scene into target and surround is useful but arbitrary because any part of a scene provides contextual information for any other part.

contralateral

On the opposite side.

contrast

The difference, usually expressed as a percentage, between the luminance (or spectral distribution, in the case of color) of two surfaces.

Cornsweet illusion

An edge effect elicited by opposing light gradients that meet along a boundary (sometimes called the Craik–O'Brien–Cornsweet illusion).

corpus callosum

The large midline fiber bundle that connects the two cerebral hemispheres.

cortex

The gray matter of the cerebral hemispheres and cerebellum, where most of the neurons in the brain are located.

cortical columns (cortical modules)

Vertically organized, iterated groups of cortical neurons that process the same or similar information. Examples are ocular dominance columns and orientation columns in the primary visual cortex.

critical period

A restricted developmental period during which the nervous systems of humans or other animals are particularly sensitive to the effects of experience.

cyclopean vision

The normal sense when looking at the world with both eyes that we see it as if with a single eye.

dark adaptation

The adjustment of the sensitivity of the visual system to dim light conditions.

degree

Unit in terms of which visual space is measured; 1° is approximately the width of the thumbnail held at arm's length and covers about 0.2mm on the retina.

dendrite

A neuronal process extending from the cell body that receives synaptic input from other neurons.

depth perception

General term used to indicate the perception of distance from the observer (can be either monocular or stereoscopic).

detector

A nerve cell or other device that responds in the presence of some stimulus feature (luminance, orientation, and so on).

dichromat

A color-deficient person (or the majority of mammals) whose color vision depends on only two cone types.

dichromatic

Having only two cone types, and thus different color perceptions than normal humans.

diencephalon

Portion of the brain that lies just rostral to the brainstem; comprises the thalamus and hypothalamus.

direction

The course taken by something, such as a point moving within a frame of reference. Direction and speed define velocity.

disparity

The geometrical difference between the view of the left and right eye in animals with frontal eyes and stereoscopic depth perception.

dorsal lateral geniculate nucleus

The thalamic nucleus that relays sensory information from the retina to the cerebral cortex. Usually referred to as the lateral geniculate or just the geniculate.

dorsal stream

Pathway from the visual cortex in the occipital lobe to the parietal cortex. Thought to support attention and object location.

eccentricity

Away from the center. In vision, typically refers to the distance in degrees away from the line of sight centered on the fovea.

edge effects

Perceptual phenomena in which the qualities at an edge affect the perception of the qualities (for example, brightness of color) of the adjoining territory (or territories). See *Cornsweet illusion*.

efferent neuron

Neuron with an axon that conducts information away from the central nervous system.

electromagnetic radiation

The full spectrum of radiation in the universe, of which light comprises only a tiny portion.

electrophysiology

Study of the nervous system by means of electrical recording.

empirical

Derived from past experience, effectively by trial and error (the opposite of *analytical*).

excitatory neuron

A neuron whose activity depolarizes (excites) the target cells it contacts.

extrastriate

Referring to the regions of visual cortex that lie outside the primary (or striate) visual cortex.

extrastriate visual areas

See *extrastriate*. Includes areas V4, MT, and MST, which are taken to be particularly pertinent to the processing of a specific visual quality (for example, color in V4, and motion in MT and MST).

feature

Physical characteristic of a stimulus, object, or condition.

feature detection

The idea that the visual system (or other sensory systems) detects and represents the characteristics of stimuli and/or the objects and conditions that give rise to them.

filling in

The perceptual attribution of a property or properties to a region of visual space when the information from that space is either absent or degraded.

fixation

Looking steadily at a particular point in visual space; the fixation point is where the lines of sight from the left and right eyes intersect.

flicker-fusion frequency

The frequency at which alternating presentations of light and dark are seen as continuous light.

fovea

Area of the human retina specialized for high acuity. Contains a high density of cones and few rods. Most mammals do not have a well-defined fovea, although many have an area of central vision (called the area centralis) in which acuity is higher than in more eccentric retinal regions.

frequency

How often something occurs within a unit of time or space.

frequency distribution

Histogram or other graphical representation showing the relative frequency of occurrence of some event or other phenomenon.

frontal lobe

One of the four lobes of the brain. Includes all of the cortex that lies anterior to the central sulcus and superior to the lateral fissure.

functional brain imaging

Technique of noninvasive brain imaging that depends on the metabolic activity of the brain tissue to reveal the location of a neural function (usually refers to positron emission topography and/or functional magnetic resonance imaging).

functional magnetic resonance imaging (fMRI)

A functional imaging technique that reveals relative brain activity based on paramagnetic differences between saturated and unsaturated blood oxygen levels.

ganglia

Collections of neurons that reside in the peripheral nervous system.

ganglion cell

Neurons in autonomic ganglia. See also *retinal ganglion cells*.

gene

A hereditary unit located on a chromosome that encodes the information needed to construct a particular protein.

genetic algorithm

A computer-based scheme for simulating the evolution of artificial neural networks.

genome

The complete set of an animal's genes.

genotype

The genetic makeup of an individual organism.

geometrical illusions

Discrepancies between the measured geometry of a visual stimulus (measurements of length, angle, and so on) and the resulting perception.

gestalt (psychology)

A school of psychology founded by Max Wertheimer in the early twentieth century in which the overall qualities of a scene are taken to be determinants of its perception. *Gestalt* in German means "shape or form."

glial cell

See *neuroglial cell.*

glomeruli

Characteristic collections of neurons and their processes in the olfactory bulbs. Formed by dendrites of mitral cells and terminals of olfactory receptor cells, as well as processes from local interneurons.

gray matter

Term used to describe regions of the central nervous system rich in neuronal cell bodies. Includes the cerebral and cerebellar cortices, the nuclei of the brain, and the central portion of the spinal cord.

growth factors

Molecules that promote the survival and growth of nerve or other cells.

gyrus

(pl. *gyri*) The ridges of the infolded cerebral cortex. The valleys between these ridges are called sulci.

Hering illusion

A classical geometrical effect in which parallel lines placed on a background of radiating lines look bowed.

heuristic

A rule derived from past experience that can be used to solve a problem when an algorithm for getting the answer is not known.

In vision, such rules are often taken to be the determinants of perception.

hidden layer

Layer in an artificial neural network that lies between the input and output layers.

hierarchy

A system of higher and lower ranks. In sensory systems, the idea that neurons in the input stages of the system determine the properties of higher-order neurons.

higher-order

Neural processes and/or brain areas taken to be further removed from the input stages of a system; sometimes used as a synonym for cognitive processes.

higher-order neurons

Neurons that are relatively remote from peripheral sensory receptors or motor effectors.

hippocampus

A specialized cortical structure located in the medial portion of the temporal lobe. In humans, concerned with short-term declarative memory, among many other functions.

hue

The aspect of color sensation (brightness and saturation being the others) that refers specifically to the qualities of red, green, blue, or yellow.

hypercomplex cells

Neurons in the primary visual cortex whose receptive field properties are sensitive to the length of the stimulus. Also called end-stopped cells, and originally thought to be determined by convergent innervation from complex cells.

illuminant

A source of illumination.

illumination

The light that falls on a scene or surface.

illusions

A much-abused word that refers to discrepancies between the physically measured properties of a visual stimulus and what is actually
seen. In fact, all percepts are illusory in this sense.

image

The representation on the retina, some other detector, or in perception
of an external form and its characteristics.

image formation

The process of focusing the light rays diverging from adjacent points
on object surfaces onto another surface (such as a screen or the
retina) to form a corresponding set of points on a two-dimensional
plane.

information

The particulars gleaned when an observer (or other receiver) can
extract a signal from the background noise.

information theory

Theory of communication channel efficiency elaborated by Claude
Shannon in the late 1940s.

inhibition

Decrease in neuronal excitability or firing rate.

inhibitory neuron

A neuron whose activity decreases the likelihood that the target cells
it contacts will, in turn, be active.

innervate

Establish synaptic contact with another neuron or target cell.

innervation

Referring to the synaptic contacts made on a target cell or larger
entity such as a muscle.

input

The information supplied to a neural or other processing system.

interneuron

A neuron in the pathway between primary sensory and primary effecter neurons; more generally, a neuron that branches locally to innervate other neurons.

inverse optics problem

The impossibility of knowing the world directly through sense information because of the ambiguity of light patterns projected onto the retina.

inverse problem

The impossibility of knowing the world directly through the senses because of the conflation of information at the level of sensory receptors.

ipsilateral

On the same side.

kludge

A machine consisting of a collection of parts cobbled together in a nonengineered way to accomplish a specific purpose.

lamina

(pl. *laminae*) One of the cell layers that characterize the neocortex, hippocampus, cerebellar cortex, spinal cord, or retina.

laminated

Layered.

Land Mondrian

A collage of papers used by Edwin Land to explore lightness and color perception. Named after early-twentieth-century abstract artist Piet Mondrian, who produced many works of this sort.

lateral inhibition

Inhibitory effects extending laterally in the plane of the retina or visual cortex; widely assumed to play a major role in perceptual phenomenology.

learning

The acquisition of novel information and behavior through experience.

lens

The transparent and spherical part of the eye whose thickening or flattening under neural control allows the light rays from objects at different distances to be focused on the retina. More generally, any object that refracts light.

light

The range of wavelengths in the electromagnetic spectrum that elicits visual sensations in humans (i.e., photons that have wavelengths of about 400–700nm).

light adaptation

See *adaptation*.

lightness

The appearance of surfaces experienced as achromatic values ranging from white through grays to black. See *brightness*.

lightness constancy

The similar appearance of two or more surfaces despite differences in the overall intensity of the spectra coming from them (typically as a function of illumination). See *color constancy*.

line (linear)

A set of points connected by a common property (such as straightness); an extension in length without thickness.

line of sight

An imaginary straight line from the center of the fovea through the point of fixation.

lobes

The four major divisions of the cerebral cortex (frontal, parietal, occipital, and temporal).

long-term potentiation (LTP)

A particular kind of enhancement of synaptic strength as a result of repeated neural activity.

luminance

The physical (photometric) intensity of light returned to the eye or some other detector, adjusted for the sensitivity of the average human observer.

luminance gradients

Gradients of light intensity.

Mach bands

Perceptual bands of lightness maxima and minima that occur at the onset and offset of luminance gradients, described by physicist Ernst Mach in 1865.

machine

Any man-made device—or, more broadly, any apparatus—that accomplishes a purpose through the operation of a series of causally connected parts.

magnocellular system

Component of the primary visual pathway specialized for the perception of motion. Named because of the relatively large nerve cells involved.

mammal

An animal whose embryos develop in a uterus and whose young suckle at birth (technically, a member of the class Mammalia).

map

A systematic arrangement of information in space. In neurobiology, the ordered projection of axons from one region of the nervous system to another, by which the organization of a relatively peripheral part of the body (such as the retina) is reflected in the organization of the nervous system (such as the primary visual cortex).

mapping

The corresponding arrangement of the peripheral and central components of a sensory or motor system.

medium

In the context of vision, a substance (such as the atmosphere or water) between the observer and the object or objects in a scene.

mesopic

Light levels at which both the rod and cone systems are active.

metamers

Two light spectra that have different wavelength distributions but nonetheless elicit the same color percepts.

microelectrode

A recording device (typically made of wire or a glass tube pulled to a point and filled with an electrolyte) used to monitor electrical potentials from individual or small groups of nerve cells.

mind

The full spectrum of consciousness at any point in time. Although used frequently in everyday speech (as in "This is what I have in mind" or "My mind is a blank"), it has little scientific meaning.

modality

A category of function. For example, vision, hearing, and touch are different sensory modalities.

module

A general term used to refer to an iterated cortical unit (ocular dominance columns, orientation columns, or blobs) found in many regions of mammalian brains.

monochromatic light

Light nominally comprising a single wavelength. In practice, often a narrow band of wavelengths generated by an interference filter.

monochromats

Color-deficient individuals who have only one or no cone opsins and, therefore, have no color vision.

monocular

Pertaining to one eye.

motion

The changing position of an object defined by speed and direction in a frame of reference.

motor

Pertaining to biological movement.

motor cortex

The region of the cerebral cortex in humans and other mammals lying anterior to the central sulcus concerned with motor behavior.

motor neuron

A nerve cell that innervates skeletal or smooth muscle.

motor system

Term used to describe all the central and peripheral structures that support motor behavior.

MST/MT

Extrastriate cortical regions in the medial temporal lobe of primates specialized for motion processing.

Müller-Lyer illusion

A geometrical effect in which the length of a line terminated by arrowheads appears shorter than the same line terminated by arrow tails. First described by nineteenth-century German philosopher and sociologist F. D. Müeller-Lyer.

muscle fibers

Cells specialized to contract when their membrane potential is depolarized.

muscle spindles

Highly specialized sensory organs found in most skeletal muscles. The spindles provide mechanosensory information about muscle length.

neocortex

The six-layered cortex that covers the bulk of the cerebral hemispheres in mammals.

nerve

A collection of peripheral axons that are bundled together and travel a common route in the body.

nerve cell

Synonym for *neuron.*

neural circuit

A collection of interconnected neurons dedicated to some neural processing goal.

neural network

Typically refers to an artificial network of interconnected nodes whose connections change in strength with experience as a means of solving problems.

neural plasticity

The ability of the nervous system to change as a function of experience; typically applied to changes in the efficacy or prevalence of synaptic connections.

neural processing

A general term used to describe operations carried out by neural circuitry.

neural system

A collection of peripheral and central neural circuits dedicated to a particular function (the visual system, the auditory system, and so on).

neuroglial cell (glial cell)

Several types of non-neural cells found in the peripheral and central nervous system that carry out a variety of functions that do not directly entail signaling.

neuromuscular junction

The synapse made by a motor axon on a skeletal muscle fiber.

neuron

Cell specialized for the conduction and transmission of electrical signals in the nervous system.

neuronal receptive field

The area of the sensory periphery (such as the retina or skin) that elicits a change in the activity of a sensory neuron.

neuronal receptive fields properties

The specific response characteristics of a neuron's receptive field.

neuroscience

Study of the structure and function of the nervous system.

neurotransmitter

A chemical agent released at synapses that affects the signaling activity of the postsynaptic target cells.

neurotransmitter receptor

A molecule embedded in the membrane of a postsynaptic cells that binds neurotransmitter.

noise

Random fluctuations that obscure a signal and do not carry information.

objects

The physical entities that give rise to visual stimuli by reflecting illumination (or by emitting light if, as more rarely happens, they are themselves generators of light).

occipital cortex

Region of the cerebral cortex nearest the back of the head, containing mainly visual processing areas.

occipital lobe

The posterior of the four lobes of the human cerebral hemisphere; primarily devoted to vision.

occlusion

Blockage of the background in a visual scene by an object in the foreground.

ocular dominance columns

The iterated stripes in the primary visual cortex of some species of primates and carnivores, arising from segregated patterns of thalamic inputs representing the two eyes.

olfactory bulb

Olfactory relay station that receives axons from the nose via cranial nerve I and transmits this information via the olfactory tract to higher centers.

ontogeny

The developmental history of an individual animal. Sometimes used as a synonym for development.

opponent colors

Colors that appear as perceptual opposites around the Newton color circle (for example, red versus green or blue versus yellow).

opsins

Proteins in photoreceptors that absorb light (in humans, rhodopsin and the three specialized cone opsins).

optic chiasm

See *chiasm.*

optic disk

The region of the retina where the axons of retinal ganglion cells exit to form the optic nerve.

optic nerve

The nerve (cranial nerve II) containing the axons of retinal ganglion cells; extends from the eye to the optic chiasm.

optic radiation

Contains the axons of lateral geniculate neurons that carry visual information to the primary visual cortex.

optic tectum

The first central station in the visual pathway of many vertebrates (homologous to the superior colliculus in mammals).

optic tract

The axons of retinal ganglion cells after they have passed through the region of the optic chiasm en route to the lateral geniculate nucleus of the thalamus.

organelle

A subcellular component visible in a light or electron microscope (nucleus, ribosome, endoplasmic reticulum, and so on).

orientation

The arrangement of an object in a two- or three-dimensional space.

orientation selectivity

Describing a neuron's selective response to edges presented over a range of stimulus orientations.

orthogonal

Making a right angle with another line or surface.

parallel processing

Processing information simultaneously in different components of a sensory (or other) system.

parietal lobe

The lobe of the human brain that lies between the frontal lobe anteriorly and the occipital lobe posteriorly.

parvocellular system

Referring to the component of the primary visual pathway in primates specialized for the detection of detail and color; named because of the relatively small size of the nerve cells involved.

perception

The subjective awareness (typically taken to be conscious) of any aspect of the external or internal environment, including thoughts, feelings, and desires.

perceptual space

The organization of a perceptual quality in subjective experience.

peripheral nervous system
All the nerves and neurons that lie outside the brain and spinal cord (that is, outside the central nervous system).

perspective
The geometrical transformation of three-dimensional objects and depth relationships when projected onto a two-dimensional surface.

phenomenology
Word used to describe the observed behavior of something.

photometer
A device that measurers the physical intensity of light.

photopic
Referring to normal levels of light, in which the predominant information is provided by cones. See *scotopic*.

photopic vision
Vision at relatively high light levels.

photoreceptors
Cells in the retina specialized to absorb photons, thus generating neural signals in response to light stimuli.

phylogeny
The evolutionary history of a species or other taxonomic category.

physiological blind spot
See *blind spot*.

pigments
Substances that both absorb and reflect light.

pixel
A discrete element in a digital image.

point
The geometrical concept of a dimensionless location in space.

posterior
Toward the back. Sometimes used as a synonym for *caudal*.

power

The rate of energy generation.

primary color(s)

The four categories of hue in human color vision (red, green, blue, and yellow). Each category is defined by a unique color perception.

primary motor cortex

A major source of descending projections to motor neurons in the spinal cord and cranial nerve nuclei. Located in the precentral gyrus and essential for the voluntary control of movement.

primary sensory cortex

Any one of several cortical areas in direct receipt of the thalamic or other input for a particular sensory modality.

primary visual cortex

The region of cortex in each occipital lobe that receives axonal projections from dorsal lateral geniculate nucleus of the thalamus. Also called Brodmann's Area 17, V1 or striate cortex. (The latter comes from the prominence of layer IV in myelin-stained sections, which gives this region a striped appearance.)

primary visual pathway

Pathway from the retina via the lateral geniculate nucleus of the thalamus to the primary visual cortex; carries the information that enables visual perception.

primate

Order of mammals that includes lemurs, tarsiers, marmosets, monkeys, apes, and humans (technically, a member of this order).

probability

The likelihood of an event, usually expressed as a value from 0 (will never occur) to 1 (will always occur).

probability distribution

Probability of a variable having a particular value, typically shown graphically.

processing

A general term that refers to the neural activity underlying some function.

psychology

The study of mental processes in humans and other animals.

psychophysics

The study of mental processes by quantitative methods, typically involving reports by human subjects of the perceptions elicited by carefully measured stimuli.

range

The distance of a point in space from an observer or a measuring device.

rank

Location on a scale, often expressed as a percentile.

real world

Phrase used to convey the idea that there is a physical world even though it is directly unknowable through the senses.

receptive field

The area of a receptor surface (such as the retina or skin) whose stimulation causes a sensory nerve cell to respond by increasing or decreasing its baseline activity. See also *neuronal receptive field.*

receptor

Cell specialized to transduce physical energy into neural signals.

receptor cells

The cells in a sensory system that transduce energy from the environment into neural signals (such as photoreceptors in the retina, hair cells in the inner ear).

reflectance

The percentage of incident light reflected from a surface.

reflection

The return of light hitting a surface as a result of its failure to be absorbed or transmitted.

reflex

A stereotyped response elicited by a defined stimulus. Usually taken to be restricted to involuntary actions.

reflex arc

Referring to the circuitry that connects a sensory input to a motor output.

refraction

The altered direction and speed of light as a result of passing from one medium to another (such as from air to the substance of the cornea).

resolution

The ability to distinguish two points in space. See *acuity*.

retina

Neural component of the eye that contains the photoreceptors (rods and cones) and the initial processing circuitry for vision.

retinal disparity

The geometrical difference between the same points in the images projected on the two retinas, measured in degrees with respect to the fovea. See *disparity*.

retinal ganglion cells

The output neurons of the retina whose axons form the optic nerve.

retinal image

The image focused on the retina by the optical properties of the eye.

retinex theory

Edwin Land's algorithm for explaining color contrast and constancy.

retinotectal system

The pathway between retinal ganglion cells and the optic tectum in vertebrates such as frogs and fish.

retinotopic map

A map in which neighbor relationships at the level of the retina are maintained at higher stations in the visual system.

retinotopy

The maintenance of the neighbor relationships at progressively higher stations in the visual system.

rhodopsin

The photopigment found in vertebrate rods.

rivalry (binocular rivalry)

The unstable visual experience that occurs when the right and left eye are presented with incompatible or conflicting images.

rods

System of photoreceptors specialized to operate at low light levels.

saccades

The ballistic, conjugate eye movements that change the point of binocular foveal fixation. These normally occur at about three or four per second.

saturation

The aspect of color sensation pertaining to the perceptual distance of a color from neutrality (thus, an unsaturated color is one that approaches a neutral gray).

scale

An ordering of quantities according to their magnitudes.

scaling

Psychophysical technique for measuring the magnitude of a sensation.

scene

The arrangement of objects and their illumination with respect to the observer; gives rise to visual stimuli.

scotoma

A defect in the visual field as a result of injury or disease to some component of the primary visual pathway.

scotopic

Referring to vision in dim light, where only the rods are operative.

sensation

The subjective experience of energy impinging on an organism's sensory receptors (a word not clearly differentiated from *perception*).

sensitivity

The relative ability to respond to the energy in a sensory stimulus.

sensory

Pertaining to sensing the environment.

sensory neuron

Any neuron involved in sensory processing.

sensory stimuli

Any pattern of energy impinging on a sensory receptor sheet such as the retina, skin, or basilar membrane in the inner ear.

sensory system

All the components of the central and peripheral nervous system concerned with sensation in a modality such as vision or audition.

simple cell

A cell in visual cortex that receives direct input from the lateral geniculate nucleus of the thalamus; the center-surround receptive fields of these neurons are organized as if constructed from the characteristics of geniculate neurons.

simultaneous brightness contrast

The ability of contextual information to alter the perception of luminance (the lightness or brightness) of a visual target.

somatic sensory cortex

That region of the mammalian neocortex concerned with processing sensory information from the body surface, subcutaneous tissues, muscles, and joints; in humans, located primarily in the posterior bank of the central sulcus and on the post-central gyrus.

somatic sensory system

The components of the nervous system that process sensory information about the mechanical forces that act on both the body surface and deeper structures such as muscles and joints.

species

A taxonomic category subordinate to genus. Members of a species are defined by extensive similarities and the ability to interbreed.

specificity

Term applied to neural connections that entail specific discrimination by neurons of their targets.

spectral differences

Differences in the distribution of spectral power in a visual stimulus that give rise to perceptions of color.

spectral sensitivity

The sensitivity of a photoreceptor (or other detecting device) to light of different wavelengths.

spectrophotometer

A device for measuring the distribution of power across the spectrum of light.

spectrum

A plot of the amplitude of a stimulus such as light or sound as a function of frequency over some period of sampling time.

spinal cord

The portion of the central nervous system that extends from the lower end of the brainstem (the medulla) to the cauda equina in the lower back.

stereopsis

The special sensation of depth that results from fusion of the two views of the eyes when they regard relatively nearby objects.

strabismus

Misalignment of the two eyes (often congenital); compromises normal binocular vision unless corrected at an early age.

striate cortex

See *primary visual cortex*.

stimulus

A generic term for a pattern of light, sound, or other energy from the environment that activates sensory receptor cells.

sulcus

(pl. *sulci*) Valleys between gyral ridges that arise from infolding of the cerebral cortex.

synapse

Specialized apposition between a neuron and a target cell; transmits information by release and reception of a chemical transmitter agent.

synaptic potentials

Membrane potentials generated by the action of chemical transmitter agents.

synaptic vesicles

The organelles at synaptic endings that contain neurotransmitter agents.

temporal lobe

The lobe of the brain that lies inferior to the lateral fissure.

terminal

A presynaptic (axonal) ending.

thalamus

A collection of nuclei that forms the major component of the diencephalon. Although its functions are many, a primary role of the thalamus is to relay sensory information from the periphery to the cerebral cortex.

T-illusion

The longer appearance of a vertically oriented line compared to a horizontally oriented line of the same length.

transduction

The cellular and molecular process by which energy is converted into neural signals.

transmittance

The degree to which a substance allows light to pass through it (for example, the transmittance of the atmosphere).

trichromacy theory

The theory that human color vision generally is explained by the different properties of the three human cone types.

trichromatic

Referring to the three different cone types in the human retina that absorb long, medium, and short wavelengths of light, respectively.

trophic

The sustaining influence of one cell or tissue on another by the action of a trophic agent such as a growth factor.

tuning curve

Result of an electrophysiological test in which the receptive field properties of neurons are gauged. The maximum sensitivity (or responsiveness) is defined by the peak of the tuning curve.

unique hues

One of the four particular hues around the Newton color circle that are seen as having no admixture of another hue (unique red, green, blue, and yellow).

universal Turing machine

A computer that, using a series of steps, can solve any problem than can be formulated in logical terms.

V1

The primary visual cortex.

V2

The secondary visual cortex.

V4

Area of extrastriate cortex that is important in color vision.

variable

A measurement that can assume any value within some appropriate range.

ventral

Referring to the belly; the opposite of *dorsal*.

ventral stream

The steam of visual information directed toward the temporal lobe that is especially pertinent to object recognition.

vertebrate

An animal with a backbone (technically, a member of the subphylum Vertebrata).

vision

The process by which the visual system (eye and brain) uses information conveyed by light to generate visual perceptions and appropriate visually guided responses.

visual acuity

See *acuity*.

visual angle

The angle between two imaginary lines that extend from an observer's eye to the boundaries of an object or interval in space.

visual association cortices

The neocortex in the occipital lobe and in the adjacent regions of the parietal and temporal lobes devoted to higher-order visual processing.

visual field

The area of visual space normally seen by one or both eyes (referred to, respectively, as the monocular and binocular fields).

visual perception

The manifestation in consciousness of visual stimuli (and not, therefore, a necessary accompaniment of vision because vision often
occurs without any particular awareness of what is being seen).

visual pigment

In humans, rhodopsin or one of the three cone opsins that absorb
light and initiate the process of vision.

visual processing

Transformations (known or unknown) carried out on information
provided by retinal stimuli.

visual qualities

The descriptors of visual percepts (brightness, color, depth, form, and
motion).

visually guided responses

An observer's actions in response to visual stimuli.

wavelength

The interval between two crests or troughs in any periodic function.
For light, the standard way of measuring the energy of different photons (measuring the frequency of photon vibration is another way).

white light

Broadband light that is perceived as lacking color (as a neutral value
of gray ranging from black to white).

white matter

A general term that refers to large axon tracts in the brain and spinal
cord. The phrase derives from the fact that axonal tracts have a
whitish cast when viewed in the freshly cut material.

White's illusion

A classic brightness illusion that poses a major problem in explaining
visual processing of luminance.

Young–Helmholtz theory

Synonym for *trichromacy theory.*

Illustration credits

Chapter 1

Figure 1.1 McMahan, U. J. *Steve: Remembrances of Stephen W. Kuffler.* Sunderland, MA: Sinauer Associates, 1990.

Figure 1.2 Alan Peters, Sanford L. Palay, Henry Webster. *The Fine Structure of the Nervous System: Neurons and Their Supporting Cells, 3e.* Oxford University Press, 1991. Reprinted by permission.

Figures 1.3, 1.4, and 1.7 Purves, D., G. A. Augustine, D. Fitzpatrick, W. Hall, A. S. LaMantia, J. O. McNamara, and S. M. Williams SM. *Neuroscience,* 4th ed. Sunderland, MA: Sinauer Associates, 2008.

Figures 1.6 and 1.7A, B Purves, D., E. M. Brannon, R. Cabeza, S. A. Huettel, K. S. LaBar, M. L. Platt, and M. Woldorff. *Principles of Cognitive Neuroscience.* Sunderland, MA: Sinauer Associates, 2008.

Chapter 2

Figures 2.1 and 2.2 McMahan, U. J. *Steve: Remembrances of Stephen W. Kuffler.* Sunderland, MA: Sinauer Associates, 1990.

Figures 2.3 and 2.4 Purves, D., G. A. Augustine, D. Fitzpatrick, W. Hall, A. S. LaMantia, J. O. McNamara, and S. M. Williams. *Neuroscience,* 4th ed. Sunderland, MA: Sinauer Associates, 2008.

Chapter 3

Figures 3.4 and 3.5 Purves, D., G. A. Augustine, D. Fitzpatrick, W. Hall, A. S. LaMantia, J. O. McNamara, and S. M. Williams. *Neuroscience,* 4th ed. Sunderland, MA: Sinauer Associates, 2008.

Chapter 4

Figures 4.1, 4.3, and 4.4 Purves, D., G. A. Augustine, D. Fitzpatrick, W. Hall, A. S. LaMantia, J. O. McNamara, and S. M. Williams. *Neuroscience,* 4th ed., Sunderland, MA: Sinauer Associates, 2008.

Figure 4.5 Reprinted by permission of the publisher from *Body and Brain: A Trophic Theory of Neural Connections* by Dale Purves, p. 103, Cambridge, MA: Harvard University Press, Copyright © 1988 by the President and Fellows of Harvard College.

Chapter 5

Figures 5.2 and 5.6 Purves, D., G. A. Augustine, D. Fitzpatrick, W. Hall, A. S. LaMantia, J. O. McNamara, and S. M. Williams. *Neuroscience,* 4th ed. Sunderland, MA: Sinauer Associates, 2008.

Figures 5.3 and 5.4 Purves, D., and J. W. Lichtman. *Principles of Neural Development.* Sunderland, MA: Sinauer Associates, 1985.

Figure 5.5 Reprinted by permission of the publisher from *Body and Brain: A Trophic Theory of Neural Connections* by Dale Purves, p. 135, Cambridge, MA: Harvard University Press, Copyright © 1988 by the President and Fellows of Harvard College. (After Purves, 1986)

Figure 5.7A Hume, R. I., and D. Purves. "Geometry of Neonatal Neurones and the Regulation of Synapse Elimination." *Nature* 293 (1981): 469–471.

Figure 5.7B Hume, R. I. and Purves, D. "Apportionment of the Terminals from Single Preganglionic Axons to Target Neurones in the Rabbit Ciliary Ganglion." *The Journal of Physiology* 338 (1983): 259–275.

Figure 5.8 Purves, D., R. D. Hadley, and J. Voyvodic. "Dynamic Changes in the Dendritic Geometry of Individual Neurons Visualized over Periods of up to Three Months in the Superior Cervical Ganglion of Living Mice." *The Journal of Neuroscience* 6 (1986): 1051–1060.

Figure 5.9 Reprinted by permission of the publisher from *Body and Brain: A Trophic Theory of Neural Connections* by Dale Purves, p. 111, Cambridge, MA: Harvard University Press, Copyright © 1988 by the President and Fellows of Harvard College. (From Lichtman et al, 1987)

Chapter 6

Figure 6.1 Purves, D., G. A. Augustine, D. Fitzpatrick, W. Hall, A. S. LaMantia, J. O. McNamara, and S. M. Williams. *Neuroscience,* 4th ed. Sunderland, MA: Sinauer Associates, 2008.

Figure 6.2 Purves, D., D. Riddle, and A. LaMantia. "Iterated Patterns of Brain Circuitry (or How the Cortex Gets Its Spots)." *Trends in Neuroscience* 15 (1992): 362–368 with permission from Elsevier.

Figure 6.3 Riddle, D. R., G. Gutierrez, D. Zheng D, L. E. White, A. Richards, and D. Purves. "Differential Metabolic and Electrical Activity in the Somatic Sensory Cortex of Juvenile and Adult Rats." *The Journal of Neuroscience* 13 (1993): 4193–4213.

Figure 6.4 Purves, D., and R. B. Lotto. *Why We See What We Do: An Empirical Theory of Vision,* Sunderland, MA: Sinauer Associates, 2003. Modified from Andrews T.J., White L.E., Binder D., Purves D. "Temporal events in cyclopean vision." *Proceedings of the National Academy of Sciences* 93 (1996): 3689-3692.

Chapter 7

Figures 7.1, 7.3, and 7.8 Purves, D., and R. B. Lotto. *Why We See What We Do: An Empirical Theory of Vision.* Sunderland, MA: Sinauer Associates, 2003.

Figures 7.4 and 7.5 Purves, D., G. A. Augustine, D. Fitzpatrick, W. Hall, A. S. LaMantia, J. O. McNamara, and S. M. Williams. *Neuroscience,* 4th ed. Sunderland, MA: Sinauer Associates, 2008.

Figure 7.6 Purves, D., and J. W. Lichtman. *Principles of Neural Development.* Sunderland, MA: Sinauer Associates, 1985.

Figure 7.7 Hubel, David H. *Eye, Brain, and Vision.* New York: Scientific American Library, 1988.

Chapter 8

Figures 8.1, 8.2, 8.4, 8.5, 8.6, 8.7, 8.8, and 8.10 Purves D., and R. B. Lotto. *Why We See What We Do: An Empirical Theory of Vision.* Sunderland, MA: Sinauer Associates, 2003.

Figure 8.3 Williams, S. M., A. N. McCoy, and D. Purves. "An Empirical Explanation of Brightness." *Proceedings of the National Academy of Sciences* 95, no. 22 (1998): 13301–13306. Copyright © 1998 National Academy of Sciences, U.S.A.

Figure 8.9 Purves, D., A. Shimpi, and R. B. Lotto. "An Empirical Explanation of the Cornsweet Effect." *The Journal of Neuroscience* 19 (1999): 8542–8551.

Chapter 9

Figures 9.1, 9.2, 9.3, 9.4, 9.7, 9.8, and 9.9 Purves, D., and R. B. Lotto. *Why We See What We Do: An Empirical Theory of Vision.* Sunderland, MA: Sinauer Associates, 2003.

Figure 9.6 Purves, D., R. B. Lotto, S. M. Williams, S. Nundy, and Z. Yang. "Why We See Things the Way We Do: Evidence for a Wholly Empirical Strategy of Vision." *Philosophical Transactions of the Royal Society of London Series B*, 356 (2001): 285–297.

Figure 9.7 Lotto, R. B., and D. Purves. "An Empirical Explanation of Color Contrast." *Proceedings of the National Academy of Sciences* 97, no. 23 (2000): 12834–12839. Copyright © 2000 National Academy of Sciences, U.S.A.

Chapter 10

Figures 10.1 and 10.3 Purves, D., and R. B. Lotto. *Why We See What We Do: An Empirical Theory of Vision.* Sunderland, MA: Sinauer Associates, 2003.

Figure 10.2 Nundy, S., and D. Purves. "A Probabilistic Explanation of Brightness Scaling." *Proceedings of the National Academy of Sciences* 99, no. 22 (2002): 14482–14487. Copyright © 2002 National Academy of Sciences, U.S.A.

Figures 10.4 and 10.5B Long, F., Z. Yang, and D. Purves. "Spectral Statistics in Natural Scenes Predict Hue, Saturation, and Brightness." *Proceedings of the National Academy of Sciences* 103, no. 15 (2006): 6013–6018. Copyright © 2006 National Academy of Sciences, U.S.A.

Figure 10.5A Wyszecki, G., and W. S. Stiles. *Color Science: Concepts and Methods, Quantitative Data and Formulae,* 2nd ed. New York: Wiley, 1982.

Figure 10.6 Van Hateren, J. H., and A. Van der Schaaf. "Independent Component Filters of Natural Images Compared with Simple Cells in Primary Visual Cortex." *Proceedings of the Royal Society of London Series B* 265 (1998): 359–366.

Figure 10.7 Yang, Z., and D. Purves. "The Statistical Structure of Natural Light Patterns Determines Perceived Light Intensity." *Proceedings of the National Academy of Sciences* 101, no. 23 (2004): 8745–8750. Copyright © 2004 National Academy of Sciences, U.S.A.

Chapter 11

Figure 11.2 Purves, D., and R. B. Lotto. *Why We See What We Do: An Empirical Theory of Vision.* Sunderland, MA: Sinauer Associates, 2003.

Figures 11.3–11.9, 11.10B, 11.11–11.14 Howe CQ and Purves D. *Perceiving Geometry: Geometrical Illusions Explained by Natural Scene Statistics,* New York, NY: Springer, 2005. Copyright © 2005 Springer Science+Business Media, Inc. With kind permission of Springer Science+Business Media. Figure 11.3 is from Chapter 2, pages 16 and 17 (originally Figures 2.1 and 2.2); Figure 11.4 is from Chapter 2, page 23 (originally Figure 2.4); Figure 11.5 is from Chapter 3, page 26 (originally Figure 3.1); Figure 11.6 is from Chapter 3, page 28 (originally Figure 3.2); Figure 11.7 is from Chapter 3, page 32 (originally Figure 3.6); Figure 11.8 is from Chapter 3, page 30 (originally Figure 3.4); Figure 11.9 is from Chapter 3, page 33 (originally Figure 3.7); Figure 11.10B is from Chapter 4, page 39 (originally Figure 4.2); Figure 11.11 is from Chapter 4, page 40 (originally Figure 4.3); Figure 11.12 is from Chapter 4, page 41 (originally Figure 4.4); Figure 11.13 is from Chapter 4, page 42 (originally Figure 4.5); Figure 11.14 is from Chapter 4, page 43 (originally Figure 4.6).

Figure 11.10(A) Nundy, S., B. Lotto, D. Coppola, A. Shimpi, and D. Purves D. "Why Are Angles Misperceived?" *Proceedings of the National Academy of the Sciences* 97, no. 10 (2000): 5592–5597. Copyright © 2000 National Academy of Sciences, U.S.A.

Chapter 12

Chapter 13

Acknowledgments

I am deeply grateful to all the talented collaborators mentioned in the book; to Shannon Ravenel, Josh Sanes, and Bill Wojtach for reading the manuscript and making many good suggestions; and to Kirk Jensen and Betsy Harris for their enthusiastic help during the editorial and production phases. I am also much indebted to Duke University, which has provided a superb environment for doing science.

About the author

Dale Purves is Professor of Neurobiology, Psychology and Neuroscience, and Philosophy at Duke University. He is a graduate of Yale University and Harvard Medical School. Upon completion of an internship and assistant residency at Massachusetts General Hospital, Dr. Purves was a post-doctoral fellow in the Department of Neurobiology at Harvard Medical School and subsequently in the Department of Biophysics at University College London. He joined the faculty at Washington University School of Medicine in 1973 where he was Professor of Physiology and Biophysics, and came to Duke in 1990 as the founding chair of the Department of Neurobiology in the School of Medicine. From 2003 to 2009 he was Director of Duke's Center for Cognitive Neuroscience, and is now Director of the Neuroscience and Behavioral Disorders Program of the Duke-NUS Graduate Medical School in Singapore. He is a member of the National Academy of Sciences, the American Academy of Arts and Sciences, and the Institute of Medicine.

Index

A

Abney effect, 167
absorption spectra, 109, 241
acetylcholine, 41, 57
  muscle fibers' sensitivity to,
    44-45
achromatic, 241
action potential, 5, 7, 241
activity, effect on nervous
  system, 42-45
acuity, 241
adaptation, 241
Adelson, Ted, 123
Adler, Julius, 32
afferent neuron, 241
Algonquin Books of Chapel
  Hill, 94
algorithm, 241
amblyopia, 242
analytical, 242
anatomy
  defined, 54
  surface anatomy of brain,
    88-89

Andersen, Per, 43
Andrews, Tim, 100, 103
angiogenesis, 20
angles, geometry perception,
  191-197
aperture problem, 242
apertures
  defined, 242
  occluding apertures, 204,
    208-215
artificial intelligence, 242
artificial neural networks, 242.
  See also neural networks
association areas, response to
  faces, 119
association cortex, 242
autonomic ganglia, 55-57, 242
autonomic nervous system,
  56-58. *See also* central
  nervous system
  cellular structure of, 4
  components of, 55
  defined, 242
  experience, effect of, 42-45

invertebrate nervous
  systems, study of, 28-34
loss of synapses, effect of,
  63-65
neural specificity, 60
    chemoaffinity hypothesis,
    61-63
    electrical recordings
    from neurons, 62
    innervation in
    sympathetic ganglia,
    60-61
pharmacology of, 57
synaptic connectivity, 77-78
    balance in, 79-80
    further research in, 85
    over time, 81-84
target-dependent neuronal
  death, 71-72
axons, 4-5, 7. *See also* action
potential
  regrowth, 63

B

Babbage, Charles, 223-224
background, defined, 243
Bacon, Francis, 119
Barlow, Horace, 131
barrels (in cortex), 95
Bayes, Thomas, 227
Bayes's theorem, 243
Bayesian decision theory
  brain systems and, 228-230
  defined, 243

Baylor, Denis, 26, 30, 33
behavior
  effect of perception and
    sensory systems on,
    228-230
  as reflexive, 230-231
Benzer, Seymour, 32
Berkeley, George,
  120-121, 179
Betz, Bill, 39
Bezold–Brücke effect, 167
binding problem, 119, 243
binocular, 243
binocular rivalry, 268
biochemistry of synaptic
  transmission, 10-12, 38
Blaustein, Mordy, 52
blind spot, 243
Bliss, Tim, 43
blobs, 243
blood vessels, inhibiting
  growth of, 20
Boston's Children's
  Hospital, 20
bottom-up, defined, 243
bottom-up processing, 243
brain
  cellular structure of, 4
  defined, 244
  organizational structure of,
    12-15
  surface anatomy of, 88-89
Brain and Visual Perception
  (Hubel and Wiesel), 131

brain modules, 90-92
  over time, 92-93
  somatic sensory system,
    95-96
brain systems
  Bayesian decision theory
    and, 228-230
  computational theory and,
    223-228
  framework for, 221-222
  neural signaling in, 219-223
  study of, versus study of
    neurons, 54-55
brainstem, 244
Brenner, Sydney, 32
Brigham, 17, 19
brightness
  defined, 244
  explained, 164
  lightness versus, 124-125
  perception of
      *disconnect from
      luminance values,
      123-126, 129-139, 142
      organization of
      perception of, 161-169,
      172-178*
Brodmann, Korbinian, 13-14
Brodmann's Area, 17, 244
Brown, Michael, 79
Bueker, Elmer, 74
Burton, Harold, 53

C

Cajál, Santiago Ramon y, 4, 79
calcarine sulcus, 107, 244

cancer, angiogenesis, 20
Cannon, Walter, 58
cataracts, 114
catecholamines, 57
cell body, 244
cells, 244
cellular structure of brain and
  nervous system, 4
central nervous system.
  See also autonomic
  nervous system
  components of, 13
  defined, 244
  in "simple" creatures, 30-31
central sulcus, 244
cerebral cortex, 244
cerebrum, 245
channels, 245
chemical synaptic
  transmission, 10-12, 38
chemoaffinity hypothesis,
  61-63
chiasm, 245
circuitry, 245
clinical psychiatry in 1960s,
  18-19
clinical-pathological
  correlations in brain
  anatomy, 14
cognition, 245
cognitive neuroscience, 87, 98
Cohen, Stanley, 74
Cold Spring Harbor
  Laboratories, 87
color, defined, 245
color addition, 108

color blind, 245

color constancy
    color perception and,
        146-150
    defined, 245
    explained, 155-156

color contrast
    defined, 245
    explained, 153-156
    stimulus, 147, 150

color deficient, 246

color opponent, 246

color perception, 108-110,
    143-146, 149-160
    advantages of, 144
    color constancy, 146-149
    colored lights and, 156-158
    as empirical theory, 151-155
        criticism of, 158-160
    organization of, 161-169,
        172-178
    relative luminance and, 157
    simultaneous color
        contrast, 146
    trichromatic theory, 145-146

color space, 246

color subtraction, 108

color-opponent cells, 246

colored lights, effect on color
    perception, 156-158

colorimetry
    defined, 246
    experiments, 166-171

communication between
    neurons
    action potential, 5-7
    biochemistry of, 10-12
    synapses, 7-10

competition, 246

complementary colors, 246

computational theory, brain
    systems and, 223-228

cone opsins, 246

cones
    defined, 246
    visual processing in, 105

consciousness, defined, 247

context, 247

contralateral, 113, 247

contrast, 247

Cornsweet edge effect,
    139-141, 247

Cornsweet, Tom, 139

corpus callosum, 247

cortex, 247

cortical columns, 247

cortical modularity, 91-92
    over time, 92-93
    somatic sensory system,
        95-96

cortical neurons
    experience, effect of,
        111-114
    responses of, 111

Costantin, Roy, 52

Cowan, Max, 53-54, 64, 69,
    72, 85, 94

Creutzfeldt, Otto, 39

Crick, Francis, 3, 32, 159
critical period, 247
cyclopean vision, 101, 248

D

Dale, Henry, 10
dark adaptation, 248
Daw, Nigel, 52
degree, 248
dendrites, 4-5, 248
Dennis, Mike, 39, 45
depression, 21, 86
depth perception, 248
Descartes, René, 119
detectors, 248
determinism, free will
 versus, 232-233
DeWeer, Paul, 52
dichromat, defined, 248
dichromatic, defined, 248
diencephalon, 248
direction of motion, 248. *See
 also* occluding apertures
disparity, 249
dorsal lateral geniculate
 nucleus, 249
dorsal stream, 249
du Bois-Reymond, Emil, 6
Duke University, 93, 95

E

eccentricity, 249
Eccles, John, 8-11, 43, 100
edge effects, 249

efferent neuron, 249
electrical recordings from
 neurons, 62
electrical signals between
 neurons
 action potential, 5-7
 biochemistry of, 10-12
 synapses, 7-10
electromagnetic
 radiation, 249
electrophysiology, 249
empirical, defined, 129, 249
empirical information, visual
 perception and, 129-139
empirical theory, color
 perception as, 151-155
 criticism of, 158-160
*An Essay Toward a
 New Theory of Vision*
 (Berkeley), 120
excitatory neuron, 249
experience
 cortical neurons, effect on,
 111-114
 nervous system, effect on,
 42-45
 perception and sensory
 systems, effect on, 228-230
 scene analysis as proxy
 for, 176
 visual perception, role in,
 123-126, 129-139, 142
extrafusal muscle fibers, 49
extrastriate, defined, 250
extrastriate cortex, 107, 109

extrastriate visual cortical
areas, 108, 250
eyedness
flicker-fusion frequency
and, 101
neural basis for, 99-101

F

faces, neurological response
to, 119
Fatt, Paul, 38
feature detection, 250
features, 250
Fechner, Gustav, 108, 161
fibrillation, 44-45
filling in, 250
Fischbach, Gerry, 85
fixation, 250
flash-drag effect, 205
flash-lag effect, 204-208
Flexner, Abraham, 51
flicker-fusion frequency,
101-103, 250
fMRI (functional magnetic
resonance imaging), 251
Folkman, Judah, 20-21
fovea, 250
free will, determinism
versus, 232-233
frequency, defined, 250
frequency distribution, 251
frequency of occurrence
experiment, geometry
perception, 182-184
Friend, Dale, 17

frontal lobe, 251
frog, visual system of, 63
Fröhlich effect, 205
functional imaging, 251
functional magnetic
resonance imaging
(fMRI), 251
Furshpan, Ed, 1-2, 5-8, 11, 31

G

Galvani, Luigi, 6
ganglia
autonomic ganglia, 55-56
defined, 251
sympathetic ganglia,
innervation in, 60-61
vertebrate autonomic
ganglia, 39
ganglion cells, 251
Gaskell, Walter, 56
Gelb, Adhemar, 125
genes, 251
genetic algorithm, 251
genome, 251
genotype, 251
geometrical illusions, 252
geometry perception,
179-196, 199
angle estimations, 191-195
frequency of occurrence
experiment, 182-184
inverse problem and,
179-180
line length experiment,
184-190

gestalt
  defined, 252
  visual perception and, 131
glial cells, 29, 252
glomeruli
  defined, 252
  olfactory glomeruli, 92
Golgi, Camillo, 4
gradients, 138
grandmother cells, 118
gray matter, 252
Gregory, Richard, 196
growth factors, 252
Gutierrez, Gabriel, 96, 99
gyrus, 252

H

Hall, Zach, 31
Hamburger, Viktor, 64, 66, 69-77
handedness, neural basis for, 99
Harris, John, 39
Harvard
  clinical psychiatry in 1960s, 18-19
  neuroscience research
    in 1960s, 24-35
    prior to 1960, 1-16
  neurosurgery in 1960s, 23-24
  surgery in 1960s, 19-21
Helmholtz Club, 159
Helmholtz, Hermann von, 108, 116, 120-121, 145
Helmholtz–Kohlrausch effect, 167

Hering, Ewald, 145
Hering illusion, 252
heuristic, 252
Heuser, John, 45
hidden unit, 253
hierarchy, 253
higher-order, defined, 253
higher-order cortical processing regions, 108
higher-order neurons, 253
Hill, Archibald, 8, 37
hippocampus, 43, 253
history of medical education, 51
Hodgkin, Alan, 6-8, 26-29, 46, 52-53
homeostasis, 56
hospitals, medical schools adjacent to, 51
Houghton Mifflin, 94
Howe, Catherine Qing, 180, 182, 184, 187, 189
Hubel, David, 1-2, 12-15, 28, 31, 33, 52, 54, 65, 88, 91, 98, 103, 105, 108, 110-121, 131-132, 219, 222
hue, 163, 253
Hume, David, 120
Hunt effect, 167
Hunt, Carlton, 48-54, 58-59, 64, 66, 85
Hurvich, Leo, 98, 168
Huxley, Andrew, 6-8, 11, 27, 29, 38, 46, 52-53, 219
hypercomplex cells, 253

I

ice cube model (visual
  system), 117-118
illuminant, 253
illumination, 254
illusions
  defined, 254
  visual illusions, 121, 131
image, 254
image formation, 254
information, defined, 254
information theory, 254
inhibition, 254
inhibitory neurons, 254
innervate, defined, 254
innervation
  defined, 254
  dependence on target cell
    geometry, 80-82
  in sympathetic ganglia, 60-61
input, 255
interneuron, 255
intrafusal muscle fibers, 49
inverse optics problem,
  120-121, 255
  defined, 255
  geometry perception and,
    179-180
  motion perception and,
    202-203
invertebrate nervous systems,
  study of, 28-34
ipsilateral, 113, 255

J–K

Jameson, Dorothea, 98, 168
Jansen, Jan, 79

Kandel, Eric, 33
Kant, Immanuel, 120
Katz, Bernard, 8-11, 25, 27,
  29, 33-34, 37-50, 53, 58, 65,
  100, 219
kludge, defined, 255
Kofka, Kurt, 131
Köhler, Wolfgang, 131
Kravitz, Ed, 1-2, 12, 31, 52
Krayer, Otto, 1, 12
Kuffler, Stephen, 1-2, 9-10,
  15-17, 24, 27-34, 37, 39, 46,
  48, 52-53, 65, 75, 93, 100,
  110, 114, 132, 219-220

L

LaMantia, Anthony, 89, 92, 95
lamina, 255
laminated, 255
Land Mondrians, 147, 255
Land, Edwin, 146, 149
Langley, John, 10, 56-62
laser range scanning, 182
lateral inhibition, 256
learning
  defined, 256
  effect on nervous system,
    42-45
  neurological basis for, 114
leech, central nerve cord
  of, 30

left-eyedness
   flicker-fusion frequency
      and, 101
   neural basis for, 99-101
left-handedness, neural basis
   for, 99
LeGros Clark, Wilfred, 53
lens, 256
Lettvin, Jerry, 118
Levi, Guiseppe, 73
Levi-Montalcini, Rita, 72-77
Levinthal, Cyrus, 32
Lichtman, Jeff, 64, 66, 79-80,
   84-85, 90, 94, 137
light
   color addition, 108
   colored lights, effect on color
      perception, 156-158
   defined, 256
   responses to, 111
light adaptation, 256
lightness
   brightness versus, 124-125
   defined, 256
   explained, 164
   perception of
      disconnect from
         luminance values,
         123-126, 129-139, 142
      organization of, 161-169,
         172-178
lightness constancy, 256
likelihood function, 227
Lillie, Frank, 70
limb bud ablation, 71

line length experiment,
   geometry perception,
   184-190
line of sight, 256
lines
   defined, 256
   straight lines, physical
      sources for, 190
lobes, 257
lobster, nervous system of, 31
Locke, John, 60
Lømo, Terje, 43
Long, Fuhui, 169
long-term memory, 42
long-term potentiation (LTP),
   43, 257
Lorente de Nó, Rafael, 91
loss of synapses, effect of,
   63-65
Lotto, Beau, 137, 140, 146,
   152, 154, 158, 162, 164, 173
LTP (long-term potentiation),
   43, 257
luminance
   defined, 125, 257
   disconnect from
      brightness/lightness
      perception, 123-126,
      129-139, 142
   frequency of occurrence,
      determining, 173-175
   organization of perceptual
      qualities related to, 162
   relative luminance, effect on
      color perception, 157
luminance gradients, 257

M

M.D./Ph.D. program, 65
Mach bands, 137-139, 257
Mach, Ernst, 137-139
machine vision, 223-225
The Machinery of the Brain
(Wooldridge), 22
machines, defined, 257
magnocellular system, 257
Magrassi, Lorenzo, 83
mammals, defined, 257
maps, 257
mapping, 258
Marr, David, 223-225
Massachusetts General
Hospital, 19, 23
Massachusetts Mental Health
Center, 18
Matthews, Margaret, 64
Maxwell, James Clerk,
109, 145
McCoy, Alli, 126
McCulloch, Warren, 225
McMahan, Jack, 34, 39, 55, 79
medical education, history
of, 51
medical schools, separation
from universities, 51
medicinal leech, central nerve
cord of, 30
medium, defined, 258
memory, long-term and short-
term, 42
mesopic, 258
*The Metabolic Care of the
Surgical Patient* (Moore), 19

metamers, 258
microelectrode, 258
Miledi, Ricardo, 38-40,
44-48, 58
mind, defined, 258
misperception in space
theory, 216
misperception in time
theory, 215
Missouri Medical College, 51
modality, 258
modules
in brain, 90-92
over time, 92-93
somatic sensory system,
95-96
defined, 258
Mondrian, Piet, 147
monochromatic light, 258
monochromats, 258
monocular, defined, 259
monocular deprivation, 114
Moore, Francis, 19
motion, defined, 259
motion perception, 201-217
flash-lag effect, 204-208
inverse problem and,
202-203
misperception in space
theory, 216
misperception in time
theory, 215
occluding apertures, 208-215
motor, defined, 259
motor cortex, 259
motor neuron, 259

motor reflexes, final common pathway for actions, 100
motor system, 259
Mountcastle, Vernon, 91, 219
MST/MT, defined, 259
Müller-Lyer illusion, 259
Mumford, David, 172
muscle fibers
  defined, 259
  fibrillation, 44-45
  sensitivity to acetylcholine, 44-45
  stretch receptors in, 48-49
muscle spindles, 259
muscles, signaling from neurons, 7-10
  biochemistry of, 10-12
myelin, 107

N

Neher, Erwin, 46-47
neocortex, 260
nerves, defined, 260
nerve cells. *See* neurons
nerve growth factor, discovery of, 72-75
nervous system, 56-58
  cellular structure of, 4
  central nervous system
    components of, 13
    defined, 244
    in "simple" creatures, 30-31
  components of, 55
  defined, 242

experience, effect of, 42-45
invertebrate nervous systems, study of, 28-34
loss of synapses, effect of, 63-65
neural specificity, 60
  chemoaffinity hypothesis, 61-63
  electrical recordings from neurons, 62
  innervation in sympathetic ganglia, 60-61
pharmacology of, 57
synaptic connectivity, 77-78
  balance in, 79-80
  further research in, 85
  over time, 81-84
target-dependent neuronal death, 71-72
neural circuit, 260
neural networks, 225-227, 242, 260
neural plasticity, 260
neural processing, 260
neural signaling in brain systems, 219-223
neural specificity, 60
  chemoaffinity hypothesis, 61-63
  electrical recordings from neurons, 62
  innervation in sympathetic ganglia, 60-61
neural system, 260

neuroanatomy, 88-89
  organizational structure of
    brain, 12-15
neuroglial cell, 260
neuromuscular junction, 260
neuronal anatomy, 4
neuronal receptive field, 261
neuronal receptive fields
  properties, 261
neurons
  communication between
    action potential, 5, 7
    biochemistry of, 10-12
    synapses, 7-10
  defined, 261
  discreteness of, 4
  features of, 5
  innervation dependence on
    target cell geometry, 80-82
  loss of synapses, effect of,
    63-65
  receptive field properties, 15
  response to faces, 119
  structural polarization, 4
  study of, versus study of
    brain systems, 54-55
  target-dependent neuronal
    death, 71-72
neuroscience
  background
    prior to 1960, 1-16
    in 1960s, 24-35
    in 1970s, 37-50
  defined, 261
  focus of study, neurons
    versus brain systems, 54-55

neurosurgery in 1960s, 23-24
neurotransmitters, 261
neurotransmitter
  receptors, 261
Newton, Isaac, 143
Nicholls, John, 24-34, 39,
  79, 98
Niedergerke, Rolf, 38
Nirenberg, Marshall, 32
Njå, Arild, 62
Nobel Prize in Physiology or
  Medicine (1986), 76
noise, 41, 261
Nundy, Shuro, 164, 166

O

object speed, visual range
  of, 201
objects, defined, 261
occipital cortex, 261
occipital lobe, 261
occluding apertures, 204,
  208-215
occlusion, 261
ocular dominance
  columns, 262
olfactory bulb, 262
olfactory glomeruli, 92
ontogeny, 262
opponent colors, 262
opsins, 262
optic chiasm, 245
optic disk, 262
optic nerve
  defined, 262
  regeneration of retinal
    axons, 62

optic radiation, 262
optic tectum, 63, 262
optic tract, 263
organelles, 263
organization of perceptual
  qualities, 161-169, 172-178
organizational structure of
  brain, 12-15
orientation
  defined, 263
  of lines, effect on line length
    perception, 184-190
orientation selectivity, 263
orthogonal, 263

P

Page, Sally, 38
pain perception, analogy with
  color perception, 143
Paintin, Audrey, 38
Palay, Sanford, 5
paraldehyde, 18
parallel processing, 263
parasympathetic component
  (autonomic nervous
  system), 56
parietal lobe, 263
parvocellular system, 263
patch-clamp electrode, 47
Pavlov, Ivan, 231
Peace Corps, 22-23
Pearlman, Alan, 52
perception, 97-104
  color perception, 143-146,
    149-160
      advantages of, 144
      color constancy, 146-149

      colored lights and,
        156-158
      as empirical theory,
        151-155, 158, 160
      relative luminance
        and, 157
      simultaneous color
        contrast, 146
      trichromatic theory,
        145-146
  defined, 131, 263
  eyedness, neural basis for,
    99-101
  geometry perception,
    179-199
      angle estimations,
        191-197
      frequency of occurrence
        experiment, 182-184
      inverse problem and, 179
      line length experiment,
        184-190
  motion perception, 201-217
      flash-lag effect, 204-208
      inverse problem and, 202
      misperception in space
        theory, 216
      misperception in time
        theory, 215
      occluding apertures,
        208-215
  organization of perceptual
    qualities, 161-169, 172-178
  pain perception, analogy with
    color perception, 143
  philosophical questions
    about, 97

as reflexive, 230-231
sensory systems and, 228-230
visual perception, 108
   brightness/lightness,
    disconnect from
    luminance values,
    123-126, 129-139, 142
   color perception, 108-110
   correlation with visual
    system, 105, 108-121
   defined, 274
   flicker-fusion frequency,
    101-103
   qualities of, 124
perceptual space, 161-169,
  172-178
  for angles, 195
  defined, 263
  of line length, 187
peripheral nervous
  system, 264
peripheral targets, neuronal
  death and, 71-72
perspective, 190
  in 2-D versus 3-D object
   motion simulation, 204, 208
  defined, 264
Peter Bent Brigham Hospital,
  17, 19
pharmacology
  of autonomic nervous
   system, 57
  of neurotransmitters, 11-12
  psychoactive drug
   research, 17

phenomenology, 264
philosophical questions about
  perception, 97
philosophy, scientific bias
  against, 98-99
photometers, 264
photopic, defined, 264
photopic vision, 264
photoreceptors, 264
phylogeny, 264
physical sources for straight
  lines, 190
physiological blind spot, 243
physiology, 54
pigments
  color subtraction, 108
  defined, 264
Pitts, Walter, 225
pixels, 264
planes, lines within, 190
points, 264
Polaroid Corporation, 147
posterior, defined, 264
posterior probability
  distribution, 228
potentiation, long-term, 43
Potter, David, 1-2, 5-8, 11,
  24-25, 31
Potter, Lincoln, 41
power, defined, 265
Price, Joel, 53, 99
primary colors
  defined, 265
  perception of, 145
primary motor cortex, 265

primary sensory cortex, 265
primary visual cortex (striate
  cortex), 107, 109, 265
primary visual pathway,
  105-106, 265
primate, defined, 265
prior probability
  distribution, 227
probability, 265
probability distribution, 265
processing, defined, 266
projective geometry, 179
psychiatry, clinical, in 1960s,
  18-19
psychoactive drug
  research, 17
psychology
  defined, 266
  scientific bias against, 98-99
psychophysical functions,
  161-169, 172-178
psychophysics, 266

Q–R

Raisman, Geoff, 64
Rakic, Pasko, 89
Ramo, Simon, 22
Ramo–Wooldridge
  Corporation, 22
range, defined, 266
rank, defined, 266
rats, somatic sensory
  cortex in, 96
Ravenel, Shannon, 26, 94

real world, defined, 266
receptive field, defined, 266
receptive field properties
  of neurons, 15
  of retinal ganglion cells, 133
  of visual neurons, 110-111
    *geometry perception
      and, 198*
    *visual perception and,
      117-118*
receptor cells, 266
receptors, 266
reflectance, 266
reflection, 267
reflex arc, 267
reflexes, 231-233, 267
refraction, 267
regeneration of retinal
  axons, 62
relative luminance, effect on
  color perception, 157
research grants, 59
resolution, 267
retina, 267
retinal axons, regeneration
  of, 62
retinal disparity, 267
retinal ganglion cells
  defined, 106, 267
  receptive field properties
    of, 133
retinal image, 267
retinex theory, 267
retinotectal system, 268

retinotopic map, 268
retinotopy, 268
retirement fund
  enrollment, 59
rhodopsin, 268
Riddle, David, 95-96, 99
right-eyedness
  flicker-fusion frequency
    and, 101
  neural basis for, 99-101
right-handedness, neural basis
  for, 99
rivalry, defined, 268
river blindness, 114
rods
  defined, 105, 268
  visual processing in, 105
Rosenthal, Jean, 43
Rovainen, Carl, 52
Rubin, Louis, 94

S

saccades, 268
Sakmann, Bert, 39-47, 55, 58,
  64, 79
Sanes, Josh, 76, 85
saturation, 163, 268
scale, defined, 268
scaling, defined, 268
scaling functions, 165
scene, defined, 268
scene analysis, as proxy for
  experience, 176
Scholars of the House
  program (Yale), 3

scotoma, 269
scotopic, 269
sea slugs, nervous system
  of, 31
sensation, 269
sensitivity, 269
sensory, defined, 269
sensory modalities, types
  of, 131
sensory neurons, 269
sensory stimuli, 269
sensory systems. *See also*
  perception
  defined, 269
  perception and, 228-230
  visual system
    anatomy of, 105-109
    color perception,
      143-160
    corical neurons, effect of
      experience, 111-114
    correlation with visual
      perception, 105,
      108-121
    geometry perception,
      179-196, 199
    ice cube model, 117
    motion perception,
      201-217
    in neuroscience studies,
      14-15
    organization of
      perception qualities,
      161-169, 172-178
Sherrington, Charles, 8,
  100-102, 220, 231

short-term memory, 42

signaling. *See* communication between neurons

simple cells, 269

simple systems, study of, 28-34

simultaneous color contrast, 146

simultaneous lightness/brightness contrast, 125, 269

single ion channels, measuring activity of, 47

somatic sensory cortex, 96, 269

somatic sensory system, 95-96, 270

Somers, David, 123

space, misperception in space theory, 216

species, defined, 270

specificity, 270

spectral differences, 270

spectral sensitivity, 270

spectrophotometers, 270

spectrum
   defined, 270
   white light spectra, 144

speed of objects, visual range of, 201. *See also* flash-lag effect; motion perception

Spemann, Hans, 70

Sperry, Roger, 61-63

spinal cord
   defined, 270
   innervation in sympathetic ganglia and, 60-61

Spitzer, Nick, 45

St. Louis Medical College, 51

Stent, Gunther, 32

stereopsis, 270

Stevens, Chuck, 88

Stevens, Stanley, 162

Stiles, Walter, 168-169

stimulus, 271

strabismus, 113, 271

straight lines, physical sources for, 190

stretch receptors in muscle fibers, 48-49

striate cortex (primary visual cortex), 107, 265

structural polarization, 4

Stuart, Ann, 29-30, 32

sulcus, 271

Sung, Kyongje, 204, 206, 213

superior cervical ganglion, 61

surface anatomy of brain, 88-89

surgery in 1960s, 19-21

Sweet, William, 23

sympathetic component (autonomic nervous system), 56

sympathetic ganglia, innervation in, 60-61

synapses, 7-10
  biochemistry of transmission,
    10-12
  defined, 271
  effect of loss, 63-65
  visualizing, 83-84
synaptic connectivity, 77-78
  balance in, 79-80
  further research in, 85
  neural specificity, 60-61
  over time, 81-84
synaptic excitation, 11
synaptic inhibition, 11
synaptic noise, 41, 88
synaptic plasticity, 42
synaptic potentials, 271
synaptic transmission at
  molecular level, 40
synaptic vesicles, 11, 271

T

T-illusion, 271
target cell geometry,
  innervation dependence on,
  80-82. *See also* muscles
target-dependent neuronal
  death, 71-72
temporal lobe, 271
terminal, defined, 271
Thach, Tom, 53
thalamus, 107-108
  defined, 271
  visual processing by, 110
TIAA-CREF, 59

tilt illusion, 192-193
time
  cortical modularity over,
    92-93
  misperception in time
    theory, 215
  synaptic connectivity over,
    81-84
trachoma, 114
transduction, 272
transmittance, 272
trichromacy, 110
trichromacy theory, 110, 272
trichromatic, defined, 272
trichromatic theory, 145-146
Trinkaus, John, 3
trophic, defined, 272
trophic agents, 72
  nerve growth factor,
    discovery of, 72-75
TRW, 22
tuning curve, 272
two-cell bodies, 5

U

unique colors, perception
  of, 145
unique hues, defined, 272
universal Turing machine, 272
universities, separation from
  medical schools, 51
University College London,
  neuroscience research in
  1970s, 37-50

V

V1 (primary visual cortex), 107, 272
V2, 272
V4, 273
van der Loos, Hendrik, 95
Van Essen, David, 79
variables, defined, 273
Venezuela, Peace Corps service in, 22-23
ventral, defined, 273
ventral stream, defined, 273
vertebrate, defined, 273
vertebrate autonomic ganglia, 39
Vietnam War, 21
visible body map, 95
vision, defined, 273
visual acuity, 241
visual angle, 273
visual association areas, 108
visual association cortices, 273
visual field, 273
visual illusions, 121, 131
visual neurons, receptive field properties of, 110-111
    geometry perception and, 198
    visual perception and, 117-118

visual perception, 108
    brightness/lightness, disconnect from luminance values, 123-126, 129-139, 142
    color perception, 108-110
    correlation with visual system, 105, 108-121
    defined, 274
    eyedness, neural basis for, 99-101
    flicker-fusion frequency, 101-103
    qualities of, 124
visual pigment, 274
visual processing
    Bayesian decision theory and, 228-230
    computational theory and, 223-228
    defined, 274
visual qualities, 274
visual system
    anatomy of, 105-109
    color perception, 143-146, 149-160
        advantages of, 144
        color constancy, 146-149
        colored lights and, 156-158
        as empirical theory, 151-155, 158, 160
        relative luminance and, 157

simultaneous color contrast, 146
trichromatic theory, 145-146
corical neurons, effect of experience, 111-114
correlation with visual perception, 105, 108-121
of frog, 63
geometry perception, 179-196, 199
angle estimations, 191-195
frequency of occurrence experiment, 182-184
inverse problem and, 179
line length experiment, 184-190
ice cube model, 117-118
motion perception, 201-217
flash-lag effect, 204-208
inverse problem and, 202
misperception in space theory, 216
misperception in time theory, 215
occluding apertures, 208-215
in neuroscience studies, 14-15
organization of perception qualities, 161-169, 172-178
visualizing synapses, 83-84
visually guided responses, 274
Volta, Alessandro, 6

W

Wallach, Hans, 209
Washington University (St. Louis), 48
neuroscience faculty, 52-54
School of Medicine, 51
Watson, James, 3, 32, 87
wavelength, 274
Wertheimer, Max, 131
white light, 144, 274
white matter, 274
White's effect, 173, 274
White, Len, 99, 107, 124
Wiesel, Torsten, 1-2, 12-15, 28, 31, 33, 52, 54, 65, 88, 91, 98, 103, 105, 108-121, 131-132, 219, 222
Williams, Mark, 124, 126, 137
Wojtach, Bill, 204, 206, 213
Wooldridge, Dean, 22
Woolsey, Tom, 53, 95
Wundt, Wilhelm, 108
Wyszecki, Gunter, 168-169

X–Y–Z

Yale
premed requirements, 2
Scholars of the House program, 3
Yale Medical School, 26
Yang, Zhiyong, 169, 172, 201
Young, Thomas, 108, 145
Young–Helmholtz theory, 145-146

Zemel, Richard, 172
Zöllner illusion, 192-193

The life sciences revolution is transforming
our world as profoundly as the industrial
and information revolutions did
in the last two centuries.
FT Press Science will capture the excitement and
promise of the new life sciences, bringing breakthrough
knowledge to every professional and interested citizen.
We will publish tomorrow's indispensable work in
genetics, evolution, neuroscience, medicine,
biotech, environmental science, and whatever
new fields emerge next.
We hope to help you make sense of the future,
so you can *live* it, *profit* from it, and *lead* it.